The Big Bang and God

The Big Bang and God

An Astro-Theology

wherein an astronomer and a theologian offer a study of interdisciplinary convergences with natural theology both in the scientific researches of Sir Fred Hoyle and in the philosophical researches of Charles Hartshorne and Alfred North Whitehead, thereby illustrating a constructive postmodern trend

Theodore Walker Jr. and Chandra Wickramasinghe

palgrave
macmillan

THE BIG BANG AND GOD

First published in 2015 by
PALGRAVE MACMILLAN®
in the United States—a division of St. Martin's Press LLC,
175 Fifth Avenue, New York, NY 10010.

Where this book is distributed in the UK, Europe and the rest of the world, this is by Palgrave Macmillan, a division of Macmillan Publishers Limited, registered in England, company number 785998, of Houndmills, Basingstoke, Hampshire RG21 6XS.

Palgrave Macmillan is the global academic imprint of the above companies and has companies and representatives throughout the world.

ISBN: 978–1–137–55242–6

Library of Congress Cataloging-in-Publication Data

Walker, Theodore, 1953–
The big bang and God : an astro-theology / Theodore Walker Jr. and Chandra Wickramasinghe.
pages cm
Includes bibliographical references and index.
ISBN 978–1–137–55242–6—
ISBN 1–137–55242–5
1. Astronomy—Religious aspects—Christianity. 2. Religion and science. 3. Cosmology. I. Wickramasinghe, Chandra, 1939– II. Title.

BL253.W35 2015
201′.65231—dc23 2015009996

A catalogue record of the book is available from the British Library.

Design by Newgen Knowledge Works (P) Ltd., Chennai, India.

First edition: September 2015

10 9 8 7 6 5 4 3 2 1

To Sir Fred Hoyle (1915–2001) and Charles Hartshorne (1897–2000)

Αἰωνία ἡ μνήμη

Contents

Appendices

Preface

By references to scientific literature, including many technical papers, an *interdisciplinary convergence* is historically described, critically evaluated, and constructively advanced. This study of scientific literature by and about Sir Fred Hoyle and his collaborators shows how *astronomy, biology, astrobiology, astrophysics, and cosmology converge with natural theology*. And, as advanced by constructive postmodern scholars instructed by mathematician-philosopher Alfred North Whitehead and logician-metaphysician-philosopher Charles Hartshorne, *natural theology yields biology and psychology with cosmological scope*. Accordingly, this book could have been titled "Convergence to God: Astronomy, Biology, Astrobiology, Astrophysics, Psychology, Cosmology, and Natural Theology."

This book is coauthored by an astronomer (astrobiologist and long-time Hoyle research collaborator Chandra Wickramasinghe) collaborating with a theologian (theological ethicist Theodore Walker Jr.). An early modern label appropriate to such interdisciplinary convergence is "Astro-Theology" (Derham 1715).

Our contemporary revision—of an early modern astro-theology—illuminates a constructive postmodern trend. At fundamental levels, advancing natural scientific disciplines are converging with natural theology.

Illuminating this constructive postmodern trend by reference to technical scientific literature (plus popular scientific writings by technical researchers) in each of the converging disciplines (astronomy, biology, astrobiology, astrophysics, psychology, cosmology, philosophy of nature, and natural theology) requires a sizable interdisciplinary bibliography. Also, this constructive postmodern trend includes natural scientific disciplines converging (or reconverging) with various artistic disciplines, especially poetry. Hence, poets (including Samuel Taylor Coleridge and

Edgar Allan Poe!), science fiction writers (including Octavia E. Butler, Arthur C. Clarke, and Fred Hoyle!), and other artists appear throughout this study. Artists are essential to the history of science, especially constructive postmodern science. And documenting this enlarges an already unavoidably large interdisciplinary bibliography.

Acknowledgments

Gratitude from Theodore Walker Jr. to, among others, the following individuals, communities, institutions, and congregations: William Abraham, Bert Affleck, Pradeep Agnihotra, Claudia J. Alexander, Darrell Alexander, Zak Alexander, Carroll A. Watkins Ali, Joseph Allen, Tracy Anne Allred, Christopher Anderson, Edwin Anderson, Javed Azam, William Babcock, Thurbert Baker, Karen Baker-Fletcher, Patricia Barrow, Roy Barrow Jr., Roy Barton, Jouette Bassler, Eugene Belcher, Herbert L. G. Bell, Marjorie Boles, Milton Boyd, William Boyd, Joseph Bracken, Brad Braxton, Lillie Brewer, Doris Faye Brinson, Leslie Brinson, Reginald Broadnax, Donna Gail Brooks, Ellis Brooks, Gloria Brooks, Cynthia Brown, Frank Brown, John Brown, Leamon Brown, Linda Brown, Robert V. "Bob" Brown, William Bryan, Peter Buffett, T. C. Buffin, Brenda Bulger-Walker, Tonya Burton, Steve Cammers, Ted Campbell, Katie Cannon, Fred Carney, Cashmere, Darrell Chambers, Poppy Chambers, Juanita Charles, Chua Ek Kay, Jaime Clark-Soles, Dawn Clayton, George Clinton, John Cobb, Jay Cole, Herbert Coleman, Jill Coleman, James H. Cone, Jennie Cook, Mary Crawford, Charles Curran, Marvel Daniels, Matrice Davis, Karen Denard, Vine Deloria Jr., my theology teacher Philip Devenish, Onais Dickson, Paula Dobbs-Wiggins, Isabel Docampo, Carolyn Douglas, Janice Drumright, Joe Drumright, Jane Elder, Fred Evans, Robin Fahmi, Martha Flowers, Clarence Ford, Patricia Francies, Mallia Franklin, Ellen Frost, George Gadda, Burke Gerstenschlager, Phil Getz, Willacin Gholston, Giggi Gilliam, Norman Girardot, Jacqueline Grant, Brenda Green, Mike Greene, David Ray Griffin, Clara Grissitt, Ruben Habito, Otis Hairston, Janelle Hampton, Sang Hun Han, Duane Harbin, Norman Harris, Richard Hartnick, Charles Hartshorne, my theological ethics teacher and dissertation director Stanley Hauerwas, Lois Patricia Hauser, Michael Hawn, Elaine Heath, Roy Heller, Linda Hervey, Theodore Hesburg, Jerome

Higgins, Perry Hodge, Zan Wesley Holmes, Kenneth Holt, Sally Hoover, Dwight Hopkins, Ricky House, Virgil Howard, Barry Hughes, Robert Hunt, James A. Jackson, Lillie Jenkins, Deepa John, Susanne Johnson, Scott Jones, William R. Jones, Rhawn Joseph, Christian Kakez-A-Kapend, A. J. Keane, Catherine Keller, Michael Kelly, Martha Kerns, Floyd Kerr, Ramon Khanna, James Kirby, Jeff LaSala, William Lawrence, Theodore and Catherine Lawson, Evelyn Leathers, Randell Leathers, Mary Leazer, Tamara Lewis, Johnny Lofton, my history of religions teacher Charles H. Long, Robin Lovin, Roger Loyd, Hugo Magallanes, David Maldonado, Robert Malloy, Natalia Marandiuc, Bruce Marshall, Mary Ann Marshall, William May, Freedom McAdoo, Cormac McCarthy, Gillian McCombs, David McDonald, William McElvaney, Neil McFarland, Timothy McGee, Timothy McLemore, Diane Miles, Rebekah Miles, Ron Miller, Debra Milton, Linn Mims, Jylla Moore, Theodora Moss, Alexis Nelson, Richard Nelson, Jessye Norman, Doug O'Dell, Schubert Ogden, Joseph Page, Mary Palmore, Evelyn Parker, Linda Patterson, Peter Paris, Rebecca Payan, Luis Pedraja, Jean Powell, Arthur Pressley, Frank Price, Marjorie Procter-Smith, Clyde Putman, Harold Recinos, Gene Reeves, Gregory Reid, Jeff Reileigh, Herbert Richardson, Joerg Rieger, Lionel Rippy, J. Deotis Roberts, Louise Marie Rountree, Daniel Dennis Sain, Donald Sandlin, Ronald Sandlin, Carolyn Santinga, Samuel Saulters, Roberta Schaafsma, Rudy Schild, Sheldon Shipman, Abraham Smith, Thomas Spann, Vera Sprunt, Lyle Staab, Roland Staton, Timothy Staveteig, Butch Steele, Faye Sturdavant, my observational astronomy teacher–(and musician) William Suggs, Ed Sylvest, Shirley Taulton, Sharon Terrell, Thandeka, Page Thomas, Howard Thurman, Jim Torrence, Darryl Trimiew, Clyde Valentin, Jason Vickers, Donald Wayne Viney, Alexander F. Vishio, Anton Vishio Jr., Allison Walker, Corey D. B. Walker, Jerry Walker, Kara Walker, Kenneth and Natalie Walker, Mary Walker, Melvin Walker, Thelmore M. and Ethel Mae Walker, my parents Theodore D. and Mary Edna Woods Walker, Sze-kar Wan, Michael Waters, Odell Watson, Thomas Webster, Anne LeVonne Baird Wells, Eric White, John Wesley White, D. Max Whitfield, Robert Wilken, Gayraud Wilmore, Charles Winslow, Charles Wood (for calling my attention to William Derham), Lewis Woodruff, Emory Barnard Woods, Zebulon Vance and Emily "Daisy" Smith Woods, Jacqueline Woodson, David Wray, Gail Yon, and Larry Yon. Many special thanks to Phil Devenish and Alex Vishio for rightly critical responses to many various drafts, and to Vishio for proofing and editing, as well as for conceptual corrections to Chapter 4. Special thanks to the Palgrave Macmillan copy editor, and the whole editorial team, for rigorous attention to a multitude of details. Also, thank you to the peer reviewers for very helpful suggestions. And gratitude goes to the administrators, faculty, librarians, libraries (especially the Perkins-SMU Bridwell Library, the SMU Fondren Library, and the online Library of Congress [also, Google Scholar]), staff, and students at Hood Theological Seminary in Salisbury, North Carolina; Bethune-Cookman College in Daytona Beach, Florida; and the Perkins School of Theology at Southern Methodist University in Dallas, Texas. And as part of gratitude to God, gratitude to churches, especially Shiloh Baptist Church in Greensboro, North Carolina; Allen Chapel African Methodist Episcopal Church in Woodsdale near Roxboro, North Carolina; and St. Luke "Community" United Methodist Church in Dallas, Texas.

Summary Preview

This book is about converging scientific disciplines. Astronomy, physics, and chemistry were beginning to converge with biology by 1946. In that year, Fred Hoyle showed that stellar processes produced heavy elements from hydrogen. See "The Synthesis of the Elements from Hydrogen" (Hoyle 1946a). Earthlings are made from heavy elements synthesized in stars. Hoyle and his collaborators thus showed that we are made of stardust. See "Synthesis of the Elements in Stars" (1957) by E. Margaret Burbidge, Geoffrey R. Burbidge, William A. Fowler, and Fred Hoyle (B^2FH). Furthermore, Hoyle and Chandra (Nalin Chandra) Wickramasinghe presented observational evidence that interstellar clouds contain organic molecules (1962, 1967, 1969a, 1969b, 1976, 1977a, 1977b, 1978, 1979b, 1980, 1983a, 1983b, 1984, 1985). Then, in a series of publications (including "On the Nature of Interstellar Grains" [1979b]), Hoyle and Wickramasinghe presented evidence that interstellar clouds contain bacteria (probably freeze-dried and, perhaps, mostly dead) and proposed that bacteria and water came to the primal Earth via comets (1978, 1978b, 1979, 1981a [1980], 1981b, 1982a, 1982b, 1985, 1986, 1987, 1988a, 1988c, 1988d). Earth was—and continues to be—seeded by comets and cometary debris (meteorites and micrometeorites). This theory of interstellar and cometary seeding is called "cometary panspermia" (1981a [1980]). Accordingly, microbiology was extended from exclusively Earth science to astronomy, astrobiology, and comet science.

Moreover, Hoyle and Wickramasinghe argued that cosmic fine-tuning is required to make biology possible (1981b). Cosmic fine-tuning exemplifies divine attributes: cosmic intelligence (omniscience), cosmic influence (omnipotence), and cosmic providence. Hence, astrobiology requires a cosmology that is consistent with theology.

Basic types of theology are identified and evaluated in *Philosophers Speak of God* (1953) by Charles Hartshorne and William L. Reese. One type of natural theology advanced by Alfred North Whitehead, and further advanced by Hartshorne, is appropriate to the work of Hoyle and Wickramasinghe: *panentheism*.

Panentheism, from *pan-en-theos*-ism, means *all-in-God*-ism. According to panentheism, *all* that is actually real (*pan*) is included *in* (*en*) *God* (*theos*). "All things are in him," says Saint Anselm of Canterbury (Williams 1995: 388). God is the all-inclusive whole of reality (Hartshorne 1973 [1967]: 7, 12, 16; Ogden 1984b: 21). And the divine whole of reality is greater than (because inclusive and transcendent of) the sum of all parts of reality. The aggregate sum is called "universe" or "cosmos," and the greater-transcendent whole is "God." God is therefore "that than which nothing greater [better or even equal] can be conceived" (St. Anselm).*

*Anselm's Latin, according to Norman Malcolm, is: *aliquid quo nihil maius cogitari possit*. And Malcolm reports that Anselm sometimes used these two alternatives to identify God: *aliquid quo maius nihil cogitari potest*, *id quo maius cogitari nequit*, and *aliquid quo maius cogitari non valet* (1960: 41). Also, see *Anselm's Discovery* (Hartshorne 1965).

As advanced in constructive postmodern thought, *panentheism* is consistent with *panpsychism* (also called "panexperientialism" [David Ray Griffin 2007: 12]) and with *panspermia*. In other words, panoramic theology is consistent with panoramic psychology and panoramic biology.

Panspermia, meaning panoramic provisioning for the possibility of newly emerging life, agrees with the theological conception of God's universal life-favoring *providence* and with the "philosophy of organism" developed in *Process and Reality: An Essay in Cosmology* (Whitehead 1927–28). And this *generic* panspermia (that agrees with a theology of providence and a cosmological philosophy of organism/biology) is factually exemplified by "cometary panspermia" (Hoyle and Wickramasinghe 1981a [1980]).

Hence, as advanced by astronomer-cosmologist Sir Fred Hoyle and astrobiologist Chandra Wickramasinghe, *astronomy, biology, astrobiology, astrophysics, and cosmology converge with natural theology*. And as advanced by constructive postmodern scholars instructed by Alfred North Whitehead and Charles Hartshorne, *natural theology yields biology and psychology with cosmological scope*.

This study corrects the prevailing view that Hoyle was unchangeably committed to atheism, and it corrects a widely shared misconception concerning relations among astronomy, cosmology, biology, and theology. In modern astronomy and cosmology, it is often conceived that while big bang cosmology implicitly supports the idea of a divine creator (required to ignite the big bang "in the beginning"), any alternative cosmology with "no boundary" (no absolute beginning) must witness against the idea of a divine creator (Hawking 1988). And, of course, it is frequently conceived that evolutionary biology witnesses against the idea of a divine creator. As a corrective, this study reveals that *both no-boundary cosmology and evolutionary biology agree with natural theology.*

CHAPTER 1

Astro-Theology and Cosmology

1.1 Modern Astro-Theology—William Derham

The word "astro-theology" has come to have various meanings (Moseley 1847; Higginson 1855; Stevens 1969; Irvin and others 2006, 2009). Its currently accrued meanings include shamanistic, ritualistic, occult, magical, mystical, astrological, and even atheistic meanings. Three centuries ago, however, astro-theology was used literally to mean *astronomy + theology*.

1.2 *Astronomy + theology* is the meaning employed by observational astronomer and Anglican clergyman William Derham (b. 1657, d. 1735). See Derham's book *Astro-Theology: Or, A Demonstration of the Being and Attributes of God, from a Survey of the Heavens* (1715).

1.3 In contrast to "ancient reckonings," Derham described math-and-telescope-assisted astronomy as "modern reckonings" (1715: 7–9). In Derham's person and work, modern astronomical reckonings plus Anglican theology produced *modern* astro-theology.

1.4 Constructive Postmodern Revisions

We are recovering and revising Derham's modern meaning of astro-theology (astronomy + theology). Along with other constructive *post-modern* revisions, we distinguish cosmology from astronomy. Hence, we now conceive astro-theology to mean astronomy *and cosmology* + theology.

1.5 Another constructive postmodern revision of modern astro-theology needs to be signaled. This revision can be signaled by replacing

"Demonstration" with "Exemplification" in Derham's subtitle—*Or, A Demonstration of the Being and Attributes of God, from a Survey of the Heavens*. Accordingly, *demonstration* of theological truth becomes *exemplification* of theological truth.

1.6 In modern science, "demonstration" implies empirical-factual proof. In constructive postmodern science, empirical-factual evidence can only illustrate or exemplify (not demonstrate or prove [nor disexemplify or disprove]) the necessary truth of natural theology (see 6.19–27). Here, "constructive postmodern" is a label for scientific inquiry that is consistent with Alfred North Whitehead's critical appreciation of modern science in *Science and the Modern World* (1925), *Process and Reality: An Essay in Cosmology* (1927–28), *Adventures of Ideas* (1933), and *Essays in Science and Philosophy* (1947).

1.6n—The term *postmodern* was used to describe Whitehead's philosophy by John B. Cobb Jr. in 1964. See Cobb's "From Crisis Theology to the Post-Modern World" (1964; also 1988). Today, many other philosophies are described as postmodern. Some postmodern philosophies trace back to developments in the visual arts and architecture during and following the 1930s. More recently, many postmodern philosophies have come to be called "deconstructive postmodern" for seeking to deconstruct modern rationalist assumptions descended from the Enlightenment. There are various postmodern philosophies (Habermas 1992; Jencks 1992; Grenz 1996; Butler 2002; Pedraja 2002; Crockett 2013). Whitehead's postmodern philosophy, however, is "radically different" from most. See *Whitehead's Radically Different Postmodern Philosophy* (2007) by David Ray Griffin. Contrary to *deconstructive* postmodern philosophies, Whiteheadian philosophy is described as a form of *constructive* postmodernism. For example, Griffin is editor of the State University of New York series in Constructive Postmodern Thought. The idea of describing Whiteheadian process thought as a form of constructive postmodernism was encouraged by Kenneth Cauthen's books *Process Ethics: A Constructive System* (1984), *Theological Biology: The Case for a New Modernism* (1991), and *Toward a New Modernism* (1997). For more about Whitehead and modern science, see *Whitehead and the Modern World: Science, Metaphysics, and Civilization* (1950) by Victor Lowe, Charles Hartshorne, and A. H. Johnson, and see *Aquinas to Whitehead: Seven Centuries of Metaphysics of Religion* (1976) by Charles E. Hartshorne. And for more about replacing demonstration with exemplification, see 9.1 of this book. Also, consider the empirical problem of the "never seen" God (Sagan 2006 [1985 Gifford lecture]: 168), and consider a constructive postmodern solution to that problem: *where anything real is seen, the all-inclusive whole of reality (God) is always seen in part, and never seen in whole.* Every nondivine observation is subject to this "observation selection effect" (Bostrom 2002). See a review of Sagan's 1985 Gifford lectures in the appendices (9.2) of this book.

1.7 Cosmology—Fred Hoyle

Sir Fred Hoyle (b. 1915, d. 2001) is a major figure in the history of astronomy and in the history of cosmology. In 1948, with Thomas Gold and Hermann Bondi, Hoyle developed "steady state" cosmology. In 1949, in BBC radio talks, Hoyle began referring to an alternative cosmology as "big bang" cosmology (Harrison 2005: 296; Longair 2006: 335; Mitton 2008: 390; Gingerich 2014: 111) (see *The Nature of the Universe: A Series of Broadcast Lectures* [1950] by Fred Hoyle). Hoyle's labels—"steady state" and "big bang"—are still widely used. In 2000, Geoffrey Burbidge, Jayant V. Narlikar, and Hoyle developed a revised steady-state cosmology called "quasi-steady-state" cosmology.

1.8 Big Bang Cosmology—Georges Edouard Lemaître

Credit for originating and developing the cosmology Hoyle labeled "big bang" goes to several individuals. Usually, the first is Belgian astronomer and priest Georges Edouard Lemaître (Abbé Lemaître) (b. 1894, d. 1966). And, along with Albert Einstein (b. 1879, d. 1955), credit often goes to Heinrich Wilhelm Matthäus Olbers (b. 1758, d. 1840), William de Sitter (b. 1872, d. 1934), Vestro Melvin Slipher (b. 1875, d. 1969), Edwin Hubble (b. 1889, d. 1953), Milton Humason (b. 1891, d. 1972), George Gamow (b. 1904, d. 1968), Arno Penzias (b. 1933), and Robert Woodrow Wilson (b. 1936).

1.8n—According to Robert Jastrow, Vestro Melvin Slipher's 1913 discovery of a dozen red-shifted galaxies was "the first hint that the universe is expanding" (1978: K1). Also, Jastrow credits William de Sitter for producing an expanding-universe solution to Einstein's equations (1992: 24). And he credits Milton Humason, an observatory janitor who taught himself astronomy, with assisting in the development of big bang cosmology. Humason worked with Edwin Hubble. Hubble measured distances, while Humason measured speeds (Jastrow 1992: 28–29).

1.9 In 1826, Heinrich Olbers showed something paradoxical. He showed that an infinite, static universe would yield a sky filled with starlight. The problem of Olbers's paradox was solved by Hubble and Humason's 1929 discovery. Inspired by Slipher's 1913 red-shift discoveries, Hubble and Humason discovered more extragalactic red shifts. These red shifts indicated that the universe is expanding rapidly. A sufficiently rapidly expanding universe would yield a dark sky such as we now see.

1.10 The Hubble–Humason 1929 discovery of universal expansion can be explained by appealing to Lemaître. He proposed that universal expansion started from a "primeval atom" (1927, 1931, 1933, 1946, 1950). Physicist and historian-of-science Helge Kragh, however, "marks the beginning of modern big bang theory" (1993: 377) with the publication of George Gamow's "Expanding Universe and the Origin of Elements" (1946). Gamow reasoned that an expanding universe would exhibit a microwave background. And a microwave background was observed by Penzias and Wilson (see "A Measurement of Excess Antenna Temperature at 4080 Mc/s" [1965]) and cosmologically interpreted by Princeton researchers Robert Dicke and P. J. E. Peebles collaborating with Peter Roll and D. T. Wilkinson (Kragh 1993: 386–87). "Only when they [Penzias and Wilson] saw a preprint of Peebles's work did they realize that the noise might be of cosmological origin" (387). The 1965 Penzias–Wilson observation confirmed the 1946 Gamow prediction. The Gamow prediction followed from the 1929 Hubble–Humason discovery. And the Hubble–Humason discovery is explainable by appealing to Lemaître's 1927 proposal.

1.10n—See Lemaître's "Un Univers homogène de masse constant et de rayon croissant rendant compte de la vitesse radiale des nébuleuses extragalactiques" ("A homogeneous Universe of constant mass and increasing radius accounting for the radial velocity of extragalactic nebulae" [1927]). See also *Discussion sur l'évolution de l'univers* (1933) by Lemaître and others, and *L'Hypothèse de l'Atome Primitive: Essai de Cosmogonie* (1946) and *The Primeval Atom: An Essay on Cosmogony* (1950) by Lemaître. Also see "The Context of Discovery: Lemaître and the Origin of the Primeval-Atom Universe" (2007) by Helge S. Kragh and Dominique Lambert. And see *The Creation of the Universe* (1961 [1952]) by George Gamow.

1.11 Big Bang Cosmology—Edgar Allan Poe

Seventy-nine years before Lemaître's 1927 proposal, a similar proposal was published. It was written by the famous poet-writer-literary-critic Edgar Allan Poe (b. 1809, d. 1849). See Poe's *Eureka: An Essay on the Material and Spiritual Universe* (1848).

1.12 Poe's 1848 book is known by more than one title. Indeed, more than one title appeared in the original edition. The 1848 book spine shows "Eureka or the Universe." Inside that book, however, the half-title page shows "Eureka: A Prose Poem." And after the preface, there is a full title: "Eureka: An Essay on the Material and Spiritual Universe" (Levine and Levine 2004: xxvi).

1.12n—See *Eureka / Edgar Allan Poe* (2004), edited with introduction, notes, and text variants by Stuart Levine and Susan F. Levine. Here, the 1848 text is reprinted in full (more than a hundred pages with 266 numbered paragraphs) and titled "Eureka: A Prose Poem."

1.13 In 1848, Poe proposed that the universe is swelling. It is "swelling" away from a "primordial Particle" (Poe 1848 [via Levine and Levine 2004: 23–25, 129n44, 142n151–53]). In 1927, Lemaître also proposed that the radius of the universe is increasing. Lemaître's universe is increasing/swelling away from a "primeval atom."

1.14 And perhaps more so than Lemaître's *primeval atom* (1927, 1931, 1946, 1950), Poe's *primordial particle* (1848) is a cosmogonic singularity. It is the "absolute extreme of Simplicity," an all-inclusive "Oneness" from which all "originally created Matter" emerged (Poe 1848 [via Levine and Levine 2004: 23]).

1.14n—Helge Kragh distinguishes religious and philosophical speculations from scientific arguments, and he describes Poe's cosmology as a "remarkable example" of "cosmogonic speculations" that "predate the scientifically argued idea" (1996: 39). Also, Kragh argues that Lemaître's primeval atom is not a singularity. He writes, "Although Lemaître's hypothesis of 1931 was the first example of a big bang universe, it was a universe evolving from a condensed material pre-universe, the primeval atom, and not from a singularity in space-time" (Kragh 1996: 54; also Kragh 1993: 373).

1.15 Moreover, according to Poe, an expanding universe will eventually cease expanding. Then, the universe will "collapse" because of "'the attraction of gravity'" (Poe 1848 [via Levine and Levine 2004: 95]). The result will be "universal agglomeration," a "final ingathering" (96–97). At this final moment, "the stellar bodies would finally be merged in one." And "[m]atter shall have returned into its original condition . . . absolute Unity" (102). Today, such a universal collapse caused by gravitational attraction is called the big crunch.

1.16 Poe's cosmogonic theory of universal expansion (big bang) included a theory of gravity-powered universal contraction (big crunch). And by appealing to "the law of periodicity," he proposed "another creation and irradiation" (Poe 1848 [via Levine and Levine 2004: 103]). Indeed, Poe described "the '*utmost conceivable expanse*' of space" as "fluctuating . . . now shrinking, now swelling" (21). This process "will be renewed forever, and forever" (103). Today, such a fluctuating universe is called an oscillating universe.

1.16n—An everlasting succession of cosmic fluctuations (Poe 1848) can be described as a succession of "cosmic epochs." The idea of a "cosmic epoch" was developed in Alfred North Whitehead's *Process and Reality: An Essay in Cosmology* (1927–28). Also, see the discussion of time as "epochal" in "Time" (1926) by Alfred North Whitehead.

1.17 Though Poe's vocabulary is different, his concepts are familiar. Here is big bang, expanding universe, contracting universe, big crunch, and oscillating universe. As Edward Harrison remarks in *Cosmology: The Science of the Universe* (2005 [1981]: 510), Poe "anticipated the expansion, collapse, and possible oscillation of the universe." Thus, concepts essential to late modern and contemporary cosmology were presented in Poe's 1848 essay.

1.17n—In their Introduction to *Eureka / Edgar Allan Poe* (2004 [Poe 1848]), editors Stuart Levine and Susan F. Levine say their scientific consultant Bruce Twarog judges that Poe's cosmology "may loosely be said to be equivalent to the 'Big Bang' theory," that Poe's "notion of the 'reciprocity' of matter and energy" is "roughly akin to the modern understanding of the relationship represented by Einstein's $e = mc^2$," and that one could plausibly claim Poe intuited "something roughly like a black hole" (xxi).

1.18 Poe's priority—in expressing concepts essential to big bang cosmology—is recognized in *The Big Bang Never Happened* (1992 [1991]) by Eric J. Lerner. Here, Lerner quotes approvingly David Heller's statement that "Poe was actually the first to propose the Big Bang" (1992 [1991]: ix; also 117).

1.18n—Also see "Eureka: Edgar Allan Poe's Big Bang Theory and the Power of Imagination" by Harry Lee Poe (2009; Metanexus-sponsored lecture).

1.19 Poe's cosmology also includes appeal to God. Poe identifies "the *Spirit Divine*" as the first/initial cause (1848 [via Levine and Levine 2004: 106]). And he says each instance of cosmogonic "swelling" is initiated by "the Divine Volition" (Poe 1848 [via Levine and Levine 2004: 21, 42, 68]). The universe swells "into existence" and subsides "into nothingness" "at every throb of the Heart Divine" (Poe 1848 [via Levine and Levine 2004: 103]).

1.20 Hoyle's Contributions to Big Bang, Steady-State, Quasi-Steady-State, Inflationary, Oscillatory, and Multiverse Cosmologies

Hoyle gave Lemaîtrean cosmology its enduring name: "big bang." And he, and others, advanced two main alternatives: steady-state cosmology and, later, quasi-steady-state cosmology.

1.21 In 1948, Hoyle, with Thomas Gold and Hermann Bondi, advanced steady-state cosmology. The cosmic state is steady because as the cosmos/universe expands, new matter is created. The creation of new matter sustains an average constant density throughout the universe.

1.21n—See "The Steady-State Theory of the Expanding Universe" (1948) by Hermann Bondi and Thomas Gold, and "A New Model for the Expanding Universe" (1948) by Fred Hoyle. Helge Kragh says "(a)lthough Hoyle's approach differed considerably from that of Bondi and Gold, the two resulting theories had so much in common that it is reasonable to speak about a single steady state theory" (1993: 394).

1.22 In 2000, Hoyle, with Geoffrey Burbidge and Jayant V. Narlikar, advanced a "quasi-steady-state" model. They were attempting to appreciate and interpret old and new observational data that the "steady-state" model could not accommodate. And they were trying to avoid increasingly complicated big bang models with many free parameters.

1.23 The quasi-steady-state model they proposed is "an oscillatory model" (Hoyle, Burbidge, and Narlikar 2000: 227). Expansion is caused by the ongoing occurrence of "compensating positive and negative forms of energy." This model explains the observed expansion of the universe "without recourse to a primordial explosion" (227).

1.23n—See "The Observations Explained in Terms of the Quasi-Steady-State Model" (Chapter 16) in *A Different Approach to Cosmology: From a Static Universe through the Big Bang towards Reality* (2000) by Fred Hoyle, Geoffrey Burbidge, and Jayant V. Narlikar.

1.24 Beyond explaining expansion, the Hoyle–Burbidge–Narlikar model explains other observed phenomena, such as the microwave background. And there are other cosmological models that can do the same.

1.24n—Another model capable of explaining expansion and microwave background is Sir Roger Penrose's "Conformal Cyclic Cosmology" (2010). Also, see "Alternative Cosmologies" (2005a) by Chandra Wickramasinghe; "An Alternative Cosmology" (2008) by Jayant V. Narlikar and Geoffrey Burbidge; "Observational Evidence Favors a Static Universe" (2011) by David F. Crawford; and "Fluid Mechanics Explains Cosmology, Dark Matter, Dark Energy, and Life" (2013) by Carl H. Gibson. Also, see Narlikar and N. C. Wickramasinghe (1968), Alfvén (1977), Alfvén and Mendis (1977), Arp, Burbidge, Hoyle, Narlikar, and N. C. Wickramasinghe (1990), and Narlikar and Burbidge (2008).

1.25 Hoyle is usually remembered mainly for contributing to the development of steady-state and quasi-steady-state cosmological models and for naming an alternative model "big bang." And though he argued against big bang models, he nevertheless made highly significant contributions to the development of contemporary astronomy and cosmology.

1.26 In *The Scientific Legacy of Fred Hoyle* (2011 [2005]) edited by Douglas Gough, John D. Barrow describes Hoyle's work on nucleosynthesis as "a foundation stone of our understanding of the universe" (98). Moreover, Hoyle was among the first to conceive of an inflationary universe (Barrow 2011 [2005]: 96 [Rees xi; Sargent 5; Narlikar 138–39]), an idea essential to contemporary big bang cosmology.

1.26n—With regard to long-range observational astronomy, in 1960 Hoyle explained an unexpected change in the *apparent* sizes of remote galaxies (Ellis 2004). With increasing distances, apparent sizes progress from small to smaller and then, unexpectantly, to larger. As Hoyle explained, such a bizarre observational phenomenon is the result of gravitational influences refocusing the light cone. Also, Hoyle pioneered the use of digital computers for modeling stellar processes (Mitton 2008: 390).

1.27 Other Hoylean contributions to science are identified by physicist and historian-of-science Helge Kragh. In "Quantum Electrodynamics" parts I and II (1939), Fred Hoyle offered one of the first "nonlocal field theories" (Kragh 1996: 164), and Hoyle and Jayant V. Narlikar anticipated the idea of "white holes, time-reversed versions of black holes" (364). Also, Hoyle and Narlikar were among the first to propose ideas about galactic superclusters and other large-scale inhomogeneities (330) and (with Geoffrey Burbidge) ideas about "bubble universes" emerging from "little big bangs" (386). These pioneering ideas were essential to subsequent big bang and multiverse cosmologies.

1.27n—In consequence of the many free parameters in big bang theory, the generic idea of a big bang can be mathematically modeled in many various ways. To be sure, there has been a continuing stream of revisionary big bang models. And throughout this stream, Hoylean cosmological terms ("big bang," "steady state," "quasi-steady-state") and Hoylean ideas (inflation, gravitational refocusing, digital computer modeling, nonlocal fields, galactic superclusters, bubble universes, little big bangs) are present, even in the most recent models. For example, see recent accounts of gravitational wave detection efforts (Ron Cowen 2014a, 2014b). (Also, in geophysics, recent evidence indicates Hoyle may have been correct in advancing T. Gold's "pore theory" that Earth's interior includes various liquids, including oil and water, and that ocean water could have been "squeezed" upward from Earth's interior [1955: 28–39; 1963: 38–46].)

1.28 Another of Fred Hoyle's many enduring contributions to science is the story of our creation from stardust. The next chapter on interdisciplinary convergences begins with that Hoyle-told narrative.

CHAPTER 2

Interdisciplinary Convergences: From Stardust to Generalized Deity

2.1 Astrochemistry

The story of our creation from stardust was first told by Sir Fred Hoyle, starting in 1946. Sir Martin Rees, Astronomer Royal and Master of Trinity College, says concerning his predecessor:

> Sir Fred Hoyle [is] . . . the person who more than anyone else explained and quantified *this wonderful story* which shows that we are linked to the stars in an even more intimate way than the astrologers think.
>
> (2011; emphasis added)

Rees's summary of Hoyle's wonderful story of our intimate linkage to the stars is: "*We are actually made of stardust*" (2011; emphasis added).

2.2 This wonderful story told by Hoyle concerns stellar physics and chemistry. Previous generations of stars were astrochemical factories. They converted hydrogen into helium and then into heavier elements, including carbon and iron. And they exploded these synthesized elements into interstellar space. These elements became parts of subsequent stars and planets. The process of stellar evolution made our carbon-based lives possible. Telling this wonderful story contributed to the development of a new convergent science: astrochemistry.

2.3 From Astrochemistry to Astrobiology

This wonderful story is narrated again by Chandra Wickramasinghe in "From Astrochemistry to Astrobiology" (2011). Wickramasinghe's retelling describes the transition from astrochemistry to astrobiology by referring to two Hoylean arguments.

2.4 First Hoylean Argument Leading to Astrobiology

The first of Hoyle's two arguments leading from astrochemistry to astrobiology was made in 1946. See Fred Hoyle's "The Synthesis of the Elements from Hydrogen" (1946a). Concerning this argument, Wickramasinghe (2011) writes:

> The logical beginning of this story goes back to Fred Hoyle's arguments of 1946... that all the chemical elements in the world heavier than H [hydrogen] and He [helium] are synthesized in nuclear reactions in stars.

In 1946, Hoyle predicted that future research would indicate that all elements heavier than helium were synthesized in stars. Subsequent astrochemical and astrophysical research proved Hoyle's prediction was correct.

2.5 Anthropic Principle

Hoyle's arguments are sometimes said to be among the first confirmed factual predictions derived from using a now-widely-discussed method. This method of reasoning focuses on identifying conditions and events required to admit the possibility of life/biology. Since obviously we are alive, those conditions and events must have occurred. And we can predict the discovery of other evidences of those conditions and events. Today, among astronomers and cosmologists, this method is frequently described as appeal to an *anthropic principle*, or as *anthropic reasoning*.

2.5n—Another Hoylean argument classified as anthropic concerns carbon resonance levels. In *Fred Hoyle's Universe* (2005), Jane Gregory writes, "In his thinking about cosmology, Hoyle had often used anthropic arguments: that is, he believed that the universe must have certain properties (among them, his predicted state of the carbon-12 nucleus) because without these properties, there would be no people around to think about such things" (327). Hoyle's C^{12} prediction has been described as an "anthropic prediction" (Barrow and Tipler 1996 [1986]: 252–54) and as an application of "the anthropic principle" (Singh 2004: 395, 487; Gregory 2005: 64; Ewart 2013). Hoyle's C^{12} resonance level prediction was confirmed by Ward Whaling and others (Longair 2011 [2005]: 105; Arnett 2011 [2005]: 22; Kragh 1996: 299). See "The 7.68-MeV State in C^{12}" (1953) by Dunbar, Pixley, Wenzel, and Whaling.

2.6 The term "anthropic" was introduced into astronomy and cosmology by Brandon Carter. See his "Large Number Coincidences and the Anthropic Principle in Cosmology" (1974).

2.6n—Carter's use of the term "anthropic" in astronomy and cosmology was taken up by others. For example, see *The Anthropic Cosmological Principle* (1986) by John D. Barrow and Frank J. Tipler, and *Cosmic Coincidences: Dark Matter, Mankind, and Anthropic Cosmology* (1989) by John Gribbin and Martin Rees. See also Brandon Carter's "The Anthropic Selection Principle and the Ultra-Darwinian Synthesis" (1993), "Anthropic Interpretation of Quantum Theory" (2004), "Microanthropic Principle for Quantum Theory" (2007), "The Significance of Numerical Coincidences in Nature" (2007), and "Classical Anthropic Everett Model: Indeterminacy in a Preordained Multiverse" (2011). Not everyone agrees that Hoyle's correct predictions are examples of anthropic reasoning. For instances, see "When Is a Prediction Anthropic? Fred Hoyle and the 7.65 MeV Carbon Resonance" (2010) by Helge S. Kragh, and Kragh's "An Anthropic Myth: Fred Hoyle's Carbon-12 Resonance Level" (2010) and "The Origin of the Modern Anthropic Principle" (2011). In "Scientific Alternatives to the Anthropic Principle" (2007 [2004]), Lee Smolin argues that anthropic reasoning "cannot yield any falsifiable predictions, and therefore cannot be a part of science" (abstract). In addition, in "Why Anthropic Reasoning Cannot Predict Lambda" (2006), Glenn D. Starkman and Roberto Trotta argue that "the anthropic principle cannot be used to explain the value of Lambda, nor, likely, any other physical parameters" (abstract). Also, in *Anthropic Bias: Observation Selection Effects in Science and Philosophy* (2002), Nick Bostrom argues persuasively for using the terms "observation selection effects" and "observation selection theory" (159f) instead of anthropic principle/reasoning.

2.7 The term "anthropic" is misleading insofar as it refers only to humans (Bostrom 2002: 6). Carter's essay is more about *bios* than *anthropos*. His "large number coincidences" (Carter 1974) concern conditions required to admit the possibility of life generally. Therefore, in astronomy and cosmology, so-called *anthropic* reasoning should be called *biotic* reasoning or *astro-biotic* reasoning (Walker 2011) or *cosmo-biotic* reasoning.

2.7n—Similarly, "Biopic Cosmology" (180–81) and a "Biopic Teleological Argument for God" (276) are presented (via a "modified Process Theology" [xix]) by Rem B. Edwards in *What Caused the Big Bang?* (2001). Edwards argues that God caused the big bang, fine-tuning, and life-providing/providential design. Yet, he notes, "most Anthropic Cosmologists" accept only those models "that exclude Divine foresight, planning, and purpose" (Edwards 2001: 181).

2.8 In philosophy, reasoning about conditions required to admit the possibility of actuality is sometimes called *transcendental* reasoning. Carter's reasoning is about the cosmological coincidences required to admit the possibility of biological actualities. Accordingly, Carter's reasoning is a form of *transcendental* cosmo-biotic reasoning.

2.8n—*Transcendental* is a technical term in philosophical metaphysics. Strictly logical *transcendental metaphysics* is sharply distinguished from all other metaphysics, including spooky/paranormal metaphysics. See "Two Views of Metaphysics" (1987b) and "What Metaphysics Is" (1970) by Charles Hartshorne. Transcendental metaphysics is often influenced by the work of philosopher Immanuel Kant (b. 1724, d. 1804). Interestingly, Kant also advanced ideas important to modern astronomy. In 1755, Kant advanced the idea that spiral nebulae are remote galaxies, not local spirals (Tyson and Goldsmith 2004: 114–15) (an idea confirmed by Hubble in 1929), and he advanced a "nebular hypothesis" of planet formation (Tyson and Goldsmith 2004: 185). Kant attributed his 1755 ideas about the existence of remote galaxies and other "core ideas" to the 1750 work of Thomas Wright (b. 1711, d. 1786) (Benson 2014: 13–15). See Immanuel Kant's *Universal Natural History and Theory of the Heavens* (2008 [originally 1755]), and Thomas Wright's *An Original Theory or New Hypothesis of the Universe* (1971 [originally 1750]).

2.9 Starting in 1946, Hoyle predicted that scientists would find evidence of heavy-element synthesis from hydrogen in stars. They did. Hoyle was instructed here by his study of stars and by his analysis of biological requirements for heavy elements, including carbon. Hoyle's astro-biotic reasoning contributed to the emergence of a new convergent scientific discipline: astrobiology.

2.9n—See Fred Hoyle's "The Synthesis of the Elements from Hydrogen" (1946a), and his "On the Formation of Heavy Elements in Stars" (1947).

2.10 Nuclear Astrophysics

Hoyle did subsequent work on stellar nuclear fusion with Margaret Burbidge, Geoffrey Burbidge, and William Fowler. See "Synthesis of the Elements in Stars" (1957) by Burbidge, Burbidge, Fowler, and Hoyle. Their findings were pivotal for our current understanding of the universe. Neil de Grasse Tyson and Donald Goldsmith write:

> Burbidge, Burbidge, Fowler, and Hoyle unified our understanding of nuclear fusion in stars with the element production visible throughout the universe.
>
> (Tyson and Goldsmith 2004: 165)

Though skeptically considered, their 1957 publication is now recognized as a turning point in scientific astronomy and cosmology.

> Their conclusions have survived decades of skeptical analysis, so their publication stands as a turning point in our knowledge of how the universe works.
>
> (Tyson and Goldsmith 2004: 165)

Similarly, Martin Rees writes, "this classic article—known to all astronomers as 'B^2FH,' the initials of its four authors—has stood the test of time" (1997: 16). (B^2FH is also called "B^2HF" [Kragh 1996: 300–4].) Furthermore, this pivotal/classic work in stellar nuclear fusion contributed to the development of another new discipline: nuclear astrophysics.

2.10n—See "Synthesis of the Elements in Stars" (1957) by E. Margaret Burbidge, Geoffrey R. Burbidge, William A. Fowler, and Fred Hoyle, and "Origin of the Elements in Stars" (1956) by Hoyle, Fowler, Burbidge, and Burbidge. Concerning the origins of nuclear astrophysics, see "Nuclear Astrophysics: The First 50 Years" (2007) by S. E. Woosley, and *Nuclear Astrophysics* (1967) by William A. Fowler. Also see Hoyle (1946, 1954), Fowler (1984), Fowler and Hoyle (1964), Hoyle and Fowler (1973), and Pasachoff and Fowler (1974).

2.11 Second Hoylean Argument Leading to Astrobiology

The second of the two arguments leading to the development of astrobiology appeared in Hoyle's book *Frontiers of Astronomy* (1955a). Here, Hoyle argued for a preplanetary origin of life. He argued that microbial life originated from preplanetary water and stardust.

2.12 Astrobiology

The convergence of astronomy and physics produced astrophysics. In Hoyle's work, the convergences of astrophysics and chemistry produced nuclear astrophysics and astrochemistry. And these astronomical convergences with biology (Hoyle 1955) contributed to the development of astrobiology.

2.12n—For more concerning Hoyle's contributions to the development of astrobiology, see "From Astrochemistry to Astrobiology" (2011) by Chandra Wickramasinghe, and "From 'Frontiers of Astronomy' to Astrobiology" (2011) by Sun Kwok.

2.13 Astrobiology was further advanced in Hoyle's *The Relation of Biology to Astronomy* (1980b). Here Hoyle drew upon work by Chandra Wickramasinghe. Perhaps the first observational evidence favoring extraterrestrial microbial life came during the 1960s (Hoyle and Wickramasinghe 1962, 1967, 1968, 1969a, 1969b). Observations and calculations by Wickramasinghe and Hoyle indicated interstellar dust includes organic molecules.

2.13n—Also, concerning interstellar organics, see Hoyle and Wickramasinghe (1976, 1977, 1979b, 1980b, 1981a, 1981b, 1982a, 1982b, 1983a, 1983b, 1984, 1985, 1986a, 1986b, 1990, 1991, 1993, 2000), Hoyle, Wickramasinghe, and Olavesen (1978), and Hoyle, Wickramasinghe, and Al-Mufti (1985).

2.14 The idea of interstellar organic molecules was widely rejected when first advanced by Hoyle and Wickramasinghe. Subsequent observational evidence has transformed this widely rejected idea into a widely accepted idea.

2.15 Similarly, Hoyle and Wickramasinghe advanced the idea that comets contain organic materials. This initially rejected idea has now become widely accepted (Chyba and Sagan 1987; Mitton 2008). Their stronger claim (that interstellar and cometary organics include bacteria and viruses), however, remains unpopular.

2.16 Comets and Biology—Cometary Panspermia

Hoyle and Wickramasinghe advanced a form of panspermia called "cometary panspermia." According to panspermia, the seeds (*sperm*-ia) of life are distributed throughout the cosmos. This idea traces back to Greek philosopher Anaxagoras (c. 510–428 BCE) (Eicher 2013: 140).

2.16n—"Until the late nineteenth century, *panspermia* meant the passage of organisms through the Earth's atmosphere, not an incidence from outside the Earth. In this form it seems to have been used first by Lazzaro Spallanzani (b. 1729, d. 1799). But almost a century before that, Francesco Redi had carried out what can be seen as a classic experiment in the subject. He had shown that maggots appear in decaying meat only when the meat is exposed to air, inferring that whatever it was that gave rise to the maggots must have traveled to the meat through the air." (Chandra Wickramasinghe 2011 [2005b]: 79)

2.17 According to *cometary* panspermia, such interstellar seeds are circulated by comets. Thus, organic materials, water, and microbial life came to Earth via comets and cometary debris trails.

2.18 The process of cometary seeding continues. Occasionally, comets slam directly into our planet. And constantly, our orbiting planet swings through trails of cometary debris, producing Earth-bound meteors, meteor showers, and meteorites, including many daily tons of micrometeorites.

2.18n—In "Incidence of Low Density Meteoroids of the Polonnaruwa-Type," N. C. Wickramasinghe and others cite J. M. C. Plane's estimation that the "average rate of meteoroid ingress from all sources (is) . . . in the range ~ 50 to 300 tonnes per day" (2013: 4). In *A Journey with Fred Hoyle: The Search for Cosmic Life* (2005a), Wickramasinghe estimates that "some 100 tonnes of cometary material reaches our planet on a daily basis" (201). Also, for more about meteoroids and comets, see Wickramasinghe's *Cosmic Dragons: Life and Death on Our Planet* (2001), and his "Microfossils in Comet Dust and Meteorites Support Panspermia" (2010). In addition, see "Comets—a Vehicle for Panspermia" by Fred Hoyle and Chandra Wickramasinghe in *Comets and the Origin of Life* (1981a [1980]), edited by C. Ponnamperuma; and Hoyle and Wickramasinghe's "Comets" (23–35) in *Proofs that Life Is Cosmic* (1982a), *Living Comets* (1985), and *Astronomical Origins of Life: Steps towards Panspermia* (2000). Finally, see "Are Microbes Currently Arriving to Earth from Space?" (2010) by Milton Wainwright, Fawaz Alshammari, and Khalid Alabri.

2.19 Interstellar Bacteria

Another widely rejected hypothesis was advanced in *The Relation of Biology to Astronomy* (Hoyle 1980b). Drawing upon Wickramasinghe's work, Hoyle argued that interstellar clouds include granular particles of bacteria. With emphasis, Hoyle wrote, "*Interstellar grains are bacteria*" (1980b: 13). Microbial life, therefore, is distributed among stars throughout the galaxy. Indeed, as Hoyle argued:

> If the grains are bacteria, and I have presented evidence which almost demands that the grains really are bacteria, then life is not confined to a particular galaxy, or to a few galaxies. Life can spread itself through the Universe.
>
> (1980b: 23)

Evidence of interstellar bacteria forced Hoyle to conclude "life did not begin on the Earth" (1980b: 21)!

2.19n—Hoyle's bibliography (1980b: 25) lists these items he co-authored with Chandra Wickramasinghe that relate directly to this conclusion: *Lifecloud* (1978c), *Diseases from Space* (1979a), "On Graphite Particles as Interstellar Grains" (1962), "Polysaccharides and the Infrared Spectra of Galactic Sources" (1977), and "On the Nature of Interstellar Grains" (1979b).

2.20 Astrobiology and Cosmology

According to Hoyle and Wickramasinghe, biology cannot remain exclusively an Earth science. The study of life is extraterrestrial (cometary), astronomical (stellar, interstellar), galactic, and cosmological. In *The Relation of Biology to Astronomy*, Hoyle comments upon the cosmological character of microbiology:

> I suspect that the cosmic quality of microbiology will seem as obvious to future generations as the Sun being the centre of our solar system seems obvious to the present generation.
>
> (1980b: 24–25)

In astrobiology and cosmology, Hoyle and Wickramasinghe advanced a not yet widely accepted idea: *life* is a "cosmic phenomenon" (1986b).

2.20n—See "The Case for Life as a Cosmic Phenomenon" (1986b) by Hoyle and N. C. Wickramasinghe in *Nature*. Also, see their *Proofs that Life Is Cosmic* (1982a) and *Is Life an Astronomical Phenomenon?* (1982b). In *The Intelligent Universe* (1984 [c. 1983]), Hoyle describes a "living Universe" (160) and concludes that "Life" is "a cosmological phenomenon, perhaps the most fundamental aspect of the Universe itself" (161).

2.21 From Astrobiology and Cosmology to Theology

Hoyle and Wickramasinghe made many initially rejected—and then increasingly accepted—contributions to science. Hence, for over four decades, they have been repeatedly infamous, then famous, then infamous and famous again.

2.22 Though Hoyle died in 2001, Wickramasinghe continues to research and publish plentifully. Their work is now becoming even more widely known. It is *not yet widely known* that, as advanced by Hoyle and Wickramasinghe, astronomy, astrophysics, astrochemistry, astrobiology, and cosmology converge with theology.

2.23 This convergence was briefly expressed in the final chapter—"Convergence to God"—of *Evolution from Space: A Theory of Cosmic Creationism* (1981b) by Fred Hoyle and Chandra Wickramasinghe. Here, they held that any possibility for life in the universe depends upon interactively fine-tuning cosmic "coupling constants" (Hoyle and Wickramasinghe 1981b: 141–43; also Hoyle 1984 [c. 1983]: 218–19). Cosmic fine-tuning that favors life exemplifies cosmic intelligence, cosmic influence, and cosmic providence. These exemplifications are divine attributes (Hesburgh 1956 [1950]; Polkinghorne 2005; Wood 2008; Bracken 2012). Hoylean–Wickramasinghean astrobiology requires a cosmology that agrees with a generalized form of theology.

2.24 From *Cosmology and Controversy: The Historical Development of Two Theories of the Universe* (1996) by Helge Kragh, we can discern that part of the controversy—concerning the choice between two types of cosmological theories (the "absolute beginning" type and the "no absolute beginning" type)—derives from a widely shared conception of relations between cosmology and theology. During the 1950s and 1960s, many cosmologists and many theologians conceived that choosing between big bang cosmologies (with absolute beginning) and steady-state cosmologies (with no absolute beginning) implied choosing between theism and atheism. It was, and still is, widely conceived that while big bang cosmologies implicitly favor theism insofar as the big bang was ignited by God ("in the [absolute] beginning"), steady-state cosmologies implicitly favor atheism insofar as there was no absolute beginning and therefore no need for divine ignition (and hence "nothing for a Creator to do" [x], avers Carl Sagan in his introduction to Stephen Hawking's *Brief History of Time* [1988], wherein Hawking develops the atheistic implication of his "no boundary" [no absolute beginning] cosmology [1988: 50, 116, 136, 140–41, 145, 148–49, 174, 185]).

2.25 Kragh rightly notes, however, that some theologians conceive that cosmologies affirming no absolute beginning imply theism, not atheism. For example, in *God and the Astronomers* (1933), William Ralph Inge conceived that the cosmological idea of an eternal universe is fully consistent with the theological idea of an eternally creative Creator and fully consistent with the need for a continuing cosmic counterbalance to entropy (Kragh 1996: 150–51). Similarly, in *Evolution and Creation* (1926), Oliver Lodge conceived of divine creativity as an eternally continuous process (Kragh 1996: 158–59).

2.26 The widely shared misconception—that the no-absolute-beginning type of cosmology is implicitly atheistic—is sometimes supported by the selected historical fact that the leading advocate of steady-state cosmology (a "no absolute beginning" type) was Fred Hoyle, and anti-religious sentiments had been publically expressed by Hoyle (Kragh 1996: 192, 195, 252–53, 322 [also see Mitton 2008: 390; Gingerich 2014: 112–14; Rubenstein 2014: 147–48; and Stenger 2014: 178–79]). Kragh writes, "When Hoyle was in his early teens he concluded that religious ideas were just fairy tales with no foundation in reality, and he never changed this simplistic atheistic view" (1996: 253).

2.27 By 1981, however, Hoyle had changed from simplistic atheism to science-based theism. Indeed, as indicated in "Convergence to God" (1981b) and in "The Concept of a Creator" (1988b) by Hoyle and Wickramasinghe, as well as in the works of those theologians identified by Kragh (William Ralph Inge, Oliver Lodge), commitments to no boundary cosmology and evolutionary biology do not imply commitments to atheism.

CHAPTER 3

Microbiology and Cometary Panspermia in Context

3.1 Origins of Microbial Life

Geochemical studies of carbon isotopes in ancient rocks show clear evidence of microbial life. Microbial life appeared suddenly on the primitive Earth about 3.9 to 4 billion years ago. How did a planet with no microbial life suddenly come to have microbial life? The existence of microbial life on Earth does not imply that it originated here. Furthermore, by any reckoning, the chance transformation from organic molecules to microbial life is exceedingly improbable. The overwhelming complexity of even the simplest microorganism indicates a cosmic origin.

3.1n—The ultimate origins of microbial life are cosmic and divine. Furthermore, according to Alfred North Whitehead's "philosophy of organism," writes David Ray Griffin, "all actual entities, being organisms, have at least some elementary capacity for self-creation as well as for receiving and exerting efficient causation" (Griffin 2008: 368). And as Griffin notes (2008: 367), in *Science and the Modern World*, Whitehead holds that *science* "is becoming the study of organisms" and *biology* "is the study of the larger organisms; whereas *physics* is the study of the smaller organisms" (1967 [1925]: 103; emphasis added). The biological origins of the "larger organisms" include influences from the "smaller organisms" studied in physics, as well as cosmic and divine influences.

3.2 Consider the sudden appearance of microbial life 3.9 to 4 billion years ago. This time period coincides with an epoch of heavy bombardment by comets. This coincidence supports the conclusion that microbes came to Earth via comets.

3.3 Recent studies of heavy-to-light water ratios (D/H ratio) provide added evidence. D/H ratio of Earth's ocean is identical to that of an important class of comets. This ratio equivalency suggests that water came to Earth via comets.

3.3n—Also, concerning water via comets, see "Source Regions and Timescales for the Delivery of Water to the Earth" (2000) by A. Morbidelli and others. Also see "Ocean-like Water in the Jupiter-family Comet 103P/Hartley 2" (2011) by Paul Hartogh and others. And more recently, the European Space Agency's *Rosetta* spacecraft detected water with higher-than-Earth-ocean-like D/H in another Jupiter-family comet. In "67P/Churyumov-Gerasimenko, a Jupiter Family Comet with a High D/H Ratio" (2014), K. Altwegg and others conclude: "Previous cometary measurements and our new finding suggest a wide range of D/H ratios in the water within Jupiter family objects and preclude the idea that this reservoir is solely composed of Earth ocean-like water" (abstract).

3.4 Modern Science versus Religion

Our ancestors from the Stone Age instinctively perceived our true cosmic origins. To find ultimate meaning, they looked to the starry vault of the heavens. The gods and goddesses were invariably placed in the heavens. After many centuries, such heavenly ideas evolved into our present-day tapestry of religious beliefs. Nevertheless, the transition from appreciating heavenly contexts to creating anthropomorphic, Earth-centered deities must be seen, in retrospect, as a retrograde motion. It moved us away from advancing a truthful understanding of our place in the universe.

3.5 With the dawn of the Renaissance, these retrograde religious views began to invite ridicule. And rationalist ridicule helped spark a bitter conflict between science and religion.

3.6 Shortly before his death, Nicolaus Copernicus (b. 1473, d. 1543) published *De revolutionibus orbium coelestium*. Oppressive churchly responses to heliocentric, Copernican astronomy marked the beginning of an epic clash. Modern science emerged from this clash between observational astronomy and church authority (Russell 1935; Kuhn 1957).

3.7 During the time of Copernicus, the Vatican endorsed a geocentric, Aristotelian cosmology. Observational astronomers producing alternative models were threatened with churchly inquisitions. Unrepentant heretics were subject to punishments, including capital punishment.

The very near-to-his-death timing of Copernicus's 1543 publication exempted him from churchly inquisitions. His successor astronomers were not exempt. Dominican astronomer Giordano Bruno (b. 1548, executed 1600) proposed that the sun was a star orbited by Earth. And he speculated that other stars were orbited by inhabited planets. For his heretical theories, Bruno was imprisoned, tortured, and finally burned at the stake by the Roman Inquisition. Mercifully, repentant astronomer Galileo Galilei (b. 1564, d. 1642) lived his final years under house arrest.

3.8 The Copernican revolution, which started with a 1543 publication, continued for centuries. A major event in the continuing Copernican revolution was the 1859 publication of *On the Origin of Species by Means of Natural Section* by Charles Darwin (b. 1809, d. 1882). Some scholars see two scientific revolutions: a Copernican revolution, followed by a Darwinian revolution. Francisco J. Ayala, however, argues for seeing "one scientific revolution" with two stages (2008: 68). First, Copernicus removed Earth from its privileged position at the physical center of the universe. Second, Darwin removed humanity from the Earth's biological center. Hence, Darwin "completed the Copernican revolution" and thus "ushered in the beginning" of "modern" science (Ayala 2008: 50, 68).

3.9 Even within the church, there were alternative conceptions of the positions of Earth and humans. Consider a historical observation by Vatican astronomer Guy Consolmagno concerning the Middle Ages. Among many medieval churchpersons, Earth was not thought to be the privileged center of creation. Instead, Earth was conceived to be at the radically underprivileged *bottom* of creation (Consolmagno 2011). The contrast between privileged center and underprivileged bottom reveals an alternative churchly cosmology. Furthermore, medieval church scholars contributed significantly to the development of early modern science. Indeed, in *Science and the Modern World* (1967 [1925]), Whitehead attributes early "faith in the possibility of science" to "medieval theology" (13).

3.9n—See also Guy Consolmagno's *The Heavens Proclaim: Astronomy and the Vatican* (2009), and *Brother Astronomer: Adventures of a Vatican Scientist* (2000).

3.10 Though it took centuries, the revolt of the astronomers (against church-supported Earth-centered cosmology) was successful. The Copernican astronomers won. The church conceded and repented. The churchly prohibition against heliocentric astronomy was dropped. Popes have apologized for churchly mistreatments of astronomers. And today, Vatican astronomers at Vatican observatories are doing fully scientific observational astronomy. Churchly oppression of observational astronomy has become churchly support for observational astronomy.

3.11 Despite Vatican repentance and support, the early modern scientific revolt against Vatican-supported geocentric cosmology has morphed into a late modern scientific revolt against religion and theology generally. Revolt against the Vatican has become revolt against all favorable references to God. In contemporary scientific publications, favorable references to God are discouraged and sometimes even prohibited. Scientific journals and book publishers normally exclude all theology, including strictly logical natural theology. And modern science *faithfully* reduces all that is real to so-called inert matter and energy. Accordingly, late modern science prohibits even the entertainment of the idea of a universal Creator inspiring creaturely activity (creativity) throughout the cosmos (creation).

3.12 The late modern tendency, then, is to view science and religion as mutually exclusive. Accordingly, religious and theological deliberations tend to be excluded from modern science.

3.13 Rather than accepting this late modern tendency, we advance a different view. A connection between the discoveries of science and some sort of generalized "religious" worldview cannot be denied. We cannot simply dismiss arguments for an all-pervasive cosmic consciousness. For example, consider the so-called coupling constants of physics. These numbers have to be arbitrarily chosen with enormous precision. Otherwise, cosmic processes would not admit the possible development of our lives. Was this "fine-tuning" a random life-favoring fluke? Or was it the deliberate act of a super intelligence?

3.14 Enter Fred Hoyle

Fred Hoyle is an iconic figure of twentieth-century astronomy. He made long-lasting and fundamental contributions to our knowledge of the cosmos. He accepted logic and mathematics as final arbiters in all matters concerning the cosmos. In the conventional sense of that word, Hoyle started his scientific career as an *atheist*. Hoyle's fidelity to evidences from logic, mathematics, and observational science, however, pushed him toward theism.

3.15 Starting from atheism, commonsense, physics, chemistry, biology, and mathematics brought Hoyle to a theistic conclusion. He famously wrote:

> A commonsense interpretation of the facts suggests that a superintellect has monkeyed with physics, as well as with chemistry and biology, and that there are no blind forces worth speaking about in nature.
>
> (1981b: 12)

He went on to note that numerical calculations put this conclusion beyond question.

> The numbers one calculates from the facts seem to me to be so overwhelming as to put this conclusion beyond question.
>
> (1981b: 12)

Hoyle's commonsense interpretation of the factual data forced him to conclude that a superintellect was exerting cosmic influences favoring the possibility of life.

3.16 The superintellect referred to here is an integral component of the universe. It is not a superintendent deity that resides outside. It is a fully integrated *cosmic intelligence.*

3.17 Hoyle maintained that the divisions among scientific disciplines—physics, chemistry, biology—were arbitrary. His career in science involved frequently crossing such boundaries. His trespasses often led to the emergence of new insights and even new scientific disciplines. Hoyle was trained as a mathematician and a theoretical physicist. When he entered astronomy in the 1940s, he was influenced by R. A. Lyttleton, a Cambridge stellar dynamicist. Lyttleton held firmly that classical physics and dynamics were most of what was needed for astronomy.

3.18 Hoyle broke with this tradition in the 1940s. He introduced nuclear physics into astronomy by pioneering the concept of nucleosynthesis in stars. In 1946, Hoyle published "The Synthesis of the Elements from Hydrogen," a ground-breaking paper. Here, he showed how chemical elements were synthesized from hydrogen by stellar nuclear reactions. He also linked the evolution of stars to nuclear reactions in their cores (Hoyle 1946a).

3.19 The main energy-producing nuclear reaction in stars converts hydrogen into helium. The further conversion of ^{4}He (helium, atomic number 4) to ^{12}C (carbon, atomic number 12) and thence to heavier elements was thought to be impossible. Hoyle, through his calculations in nuclear physics, came to a different conclusion. He argued that there *must* be a hitherto undiscovered resonance in the nucleus of ^{12}C (an unstable energy level) if ^{4}He is to burn into ^{12}C.

3.20 The argument contained in Fred Hoyle's 1946 paper has been called an anthropic argument. We now call it an astrobiotic (or transcendental astrobiotic) argument (see 2.5–8). Biological life requires that certain propositions about the universe be true. In this case, the progression of nuclear reactions in stars from He to C, N (nitrogen), O (oxygen), and beyond is required. And this progression in turn requires the existence of a certain state in the carbon nucleus. Accordingly, Hoyle predicted discovery of the required state. This state was looked for and discovered by Ward Whaling, William Fowler, and others (Longair 2011 [2005]: 105; Arnett 2011 [2005]: 22; Dunbar, Pixly, Wenzel, and Whaling 1953; Cook, Fowler, Lauritsen, and Lauritsen 1957). This discovery confirmed Hoyle's 1946 theoretical prediction.

3.21 When massive stars explode as supernovae, the chemical elements of life synthesized within them are expelled into interstellar space. Hoyle and his collaborators—Burbidge, Burbidge, and Fowler—worked through the details of these processes. They diligently compared theoretical predictions with observations at the telescope. A new scientific discipline—nuclear astrophysics—was born in October 1957 with the publication of a landmark paper (B^2FH) in the *Review of Modern Physics*.

3.21n—See "Synthesis of the Elements in Stars" (1957) by E. Margaret Burbidge, Geoffrey R. Burbidge, William A. Fowler, and Fred Hoyle. William A. Fowler (b. 1911, d. 1988), 1983 Nobel Prize winner (with Subramanyan Chandrasekhar) in physics, in his autobiographical statements, available online at Nobelprize.org (the official website of the Nobel Prize), says: "The grand concept of nucleosynthesis in stars was first definitely established by Hoyle in 1946. After Whaling's confirmation of Hoyle's ideas I became a believer and in 1954/1955 spent a sabbatical year in Cambridge, England, as a Fulbright Scholar in order to work with Hoyle. There Geoffrey and Margaret Burbidge joined us. In 1956 the Burbidges and Hoyle came to Kellogg and in 1957 our joint efforts culminated in the publication of 'Synthesis of the Elements in Stars' in which we showed that all of the elements from carbon to uranium could be produced by nuclear processes in stars starting with the hydrogen and helium produced in the big bang" (Fowler 1983; www.nobelprize.org/nobel-prizes/laureates/1983/fowler-bio.html).

3.22 From Elements to Organic Molecules and Life

The next step in progressing from elements of life to life itself involves forming organic molecules. In the 1950s, no molecules larger than simple radicals (such as CH and CN [bonds in radical chemistry]) were known to exist outside our solar system. Hoyle was convinced, however, that larger molecules (including H_2 [hydrogen], CO [carbon monoxide], and organic molecules) *must* be distributed throughout the galaxy.

3.22n—The idea of organic molecules in outer space was not taken seriously by astronomers during the 1950s. Papers Hoyle sent to journals on this topic were rejected. So he expressed this idea by writing a science fiction novel. Hoyle's *The Black Cloud* (1957) is about an interstellar cloud of organic molecules approaching Earth. The cloud is a biological entity endowed with a metabolism and intelligence.

3.23 An alternative to the idea that microbial life originated in water on planet Earth was offered in Hoyle's *Frontiers of Astronomy* (1955a). Consider the relatively miniscule volume of water on planet Earth in relation to the astronomical volume of water in the primordial cloud that formed our solar system. Obviously, it is much more likely that life originated in that primordial cloud. Rather than accepting the idea that microbial life originated on our planet, Hoyle argued for a "preplanetary origin of life" (1963 [1955]: 100).

3.23n—Hoyle's *Frontiers of Astronomy* was first printed in July 1955, reprinted in 1955 [!], printed again in 1956, twice more [!] in 1961, and a sixth printing in June 1963; plus, it was translated into Spanish and printed in Mexico in 1960 and again in 1970. It was widely read, and it helped inspire Chandra Wickramasinghe and many others to study astronomy (Martin Rees 2005: xiii; Mitton 2008: 392).

3.24 Hoyle–Wickramasinghe Collaboration Begins

Such pioneering ideas (heavy elements synthesized in stars, interstellar distribution of water and organic molecules, preplanetary origin of microbial life) were firmly in place in Hoyle's thinking when Chandra Wickramasinghe arrived in Cambridge in 1960 and began to work as his research student. Together they worked on a project to study interstellar dust. Although not originally intended to spread outside the narrow confines of conventional astronomy, the project soon began to expand uncontrollably. Interstellar dust that was hitherto thought to be comprised of inorganic material turned out to be largely carbonaceous and organic. New observations in astronomy, facilitated by space telescopes and new techniques of ultraviolet and infrared spectroscopy, showed the widespread existence of complex organic molecules in space. Through the 1970s and 1980s, it became evident that the organics in the interstellar medium, in comets, and in the most distant galaxies were much more complex than originally conceived and bore all the hallmarks of biology. At the very least, the organic building blocks of life were present in vast quantity in the cloud from which the solar system formed. A head-on clash with conventional biology was now imminent.

3.25 The holy grail of modern biology is this hypothesis: life emerged from a primordial soup of organics generated in situ from inorganic molecules on primitive Earth (Oparin 1938). A mixture of inorganic gases in the atmosphere—water, methane, carbon dioxide, hydrogen cyanide—is claimed to have been sparked by flashes of lightning and ultraviolet light to produce a trickle of organic molecules that rained down into the primitive oceans to form a primordial soup. The claim is that from this soup of organics primitive life must have evolved after millions of years.

3.25n—See *The Origin of Life* (1938) by Aleksandr Ivanovič Oparin (b. 1894, d. 1980).

3.26 The primordial soup theory gained putative empirical support from the classic studies of Miller (1953) and Miller and Urey (1959). In these studies, minute quantities of amino acids and sugars were produced by sparking mixtures of inorganic gases. Ponnamperuma and Mack (1965) later demonstrated the production of nucleotides (components of DNA) under similar conditions in the laboratory. Finally, the experiments of Sagan and Khare (1971) showed the production of amino acids from gases exposed to ultraviolet light. All such experimental triumphs were greeted as crucial steps toward understanding the origin of life on Earth, although it was never clear that the experimental conditions used in the laboratory had any relevance to those of primitive Earth. On the contrary, the early terrestrial atmosphere that is now believed to have been oxidizing would have inhibited any synthesis of organics of the type demonstrated by these studies, and organic molecules are, in any case, a very far cry from life.

3.26n—See "A Production of Amino Acids under Possible Primitive Earth Conditions" (1953) by Stanley L. Miller; "Organic Compound Synthesis on the Primitive Earth" (1959) by Stanley L. Miller and Harold C. Urey; "Nucleotide Synthesis under Possible Primitive Earth Conditions" (1965) by Cyril Ponnamperuma and Ruth Mack; and "Long-wavelength Ultraviolet Photo Production of Amino Acids on the Primitive Earth" (1971) by Carl Sagan and B. N. Khare.

3.27 A terrestrial origin of the chemical building blocks of life might have been thought plausible before it was discovered that vast quantities of biogenic organic molecules existed within interstellar clouds (Hoyle, Wickramasinghe, and Olavesen 1978; Kwok, 2009b; Sandford and others 2009). Hoyle and Wickramasinghe were among the first to present evidence of complex organic molecules in deep space (1976, 1977a, 1977b, 1978a, 1978d, 1979a, 1979b, 1980, 1981a [October 1980], 1981b, 1981c, 1982a, 1982b, 1983a, 1983b, 1984, 1985a, 1985b, 1985c, 1986a, 1986b). (In addition, see Dayal Tissa Wickramasinghe and David A. Allen [1980, 1986], and David A. Allen and Dayal Tissa Wickramasinghe [1981].)

3.28 The total amount of organic material in the galaxy in the form of organic dust and polycyclic aromatic hydrocarbons accounts for about a third of all the carbon present in interstellar space—a truly vast quantity amounting to some billions of solar masses. And the most reasonable way to comprehend the existence of such a vast quantity of complex interstellar organics is to suppose that it was generated by life itself. Whenever organic materials are found on Earth, we do not hesitate to say that biology was responsible for their formation. Indeed, 99.999 percent of all the carbonaceous and organic deposits on the Earth can only be generated via biology.

3.29 Why not apply the same argument to organics found in interstellar space? The only possible reason for not doing so is the cultural prejudice that prevails. Modern reductionism deems that life outside the Earth in any quantity is most unlikely, and thus the biogenic origin of interstellar organics would constitute an extraordinary claim. The assertion continues that extraordinary claims need extraordinary evidence to justify them. What nonsense! On the contrary, confining life to Earth is the extraordinary claim. It is dictated only by an obsolete pre-Copernican view of the world. This was the logic that drove Hoyle and Wickramasinghe to consider the merits of panspermia.

3.30 Cometary Panspermia versus Random Assembly

Panspermia goes back to classical Greece and to the Vedic traditions of ancient India. In the West, Anaxarogas in the fifth century BCE posited that the seeds of life (spermata) were widely distributed (panned) across the cosmos. At the dawn of the twentieth century, Swedish Nobel laureate Svante August Arrhenius (b. 1859, d. 1927) argued that bacterial spores can on occasion be lofted off the surfaces of Earth-like planets and propelled by the pressure of starlight across the galaxy (1903, 1908).

3.31 According to this theory, conditions for life's possibility are present everywhere. No "miracle" or miraculous random assembly process is required.

3.32 The idea of microbial life in outer space has been supported in recent years by the discovery of the extraordinary survival properties of bacteria and bacterial spores. Microorganisms, it would seem, are born to be space travelers, able to withstand the harshest rigors of space.

3.33 Fred Hoyle and Chandra Wickramasinghe argued that the arrangements of organic molecules into living forms—for example, chains of amino acids assembled into enzymes—had such miniscule probabilities that to envisage their self-assembly taking place in the oceans of the Earth is ludicrous in the extreme. In *Evolution from Space* (Hoyle and Wickramasinghe 1981b), it was calculated that the chance of arriving at the correct arrangements of amino acids into the 2000 or so enzymes crucial for life in a random trial is only one part in $10^{40,000}$ [!].

3.34 Hoyle encapsulated the essence of this calculation in his famous "Boeing 747 analogy." He wrote:

> The chance that life forms might have emerged in this way is comparable with the chance that a tornado sweeping through a junk-yard might assemble a Boeing 747 from the material therein.
>
> (1981: 105)

By comparing the chance assembly of 2000 enzymes to the chance assembly of a Boeing 747, Hoyle was illustrating the incredible improbability of both. Similarly, in *Evolution from Space: A Theory of Cosmic Creationism*, Hoyle and Wickramasinghe wrote:

> life cannot have had a random beginning. Troops of monkeys thundering away at random on typewriters could not produce the works of Shakespeare, for the practical reason that the whole observable universe is not large enough to contain the necessary monkey hordes, the necessary typewriters, and certainly the waste paper baskets required for the deposition of wrong attempts. The same is true for living material.
>
> (1981b: 148)

Theories that appeal to such highly improbable chance assemblies are scientific failures. Like the complexity of Boeing 747s and of Shakespeare's works, the complexity of life exceeds the possibility of random assembly.

3.34n—In Chapter 1—"Chance and the Universe"—in *The Intelligent Universe: A New View of Creation and Evolution* (1984 [1983]), Fred Hoyle labels appeals to such improbable chance assemblies as "junkyard mentality" (19). Richard

Dawkins disagrees. In *The Blind Watchmaker: Why the Evidence of Evolution Reveals a Universe without Design* (1987 [1986]), Dawkins argues against impossibly improbable "Boeing 747 macromutations" made by junkyard hurricanes, and in favor of more probable "Stretched DC8 macromutations...made by modifying an earlier airliner, the DC8" (234). Critics who have challenged this "Boeing 747" improbability argument have asserted without proof that a process of step-wise natural selection toward the survival of the fittest would operate so as to lead to the emergence of a *living* cell (a *living* "Stretched DC8" *from a nonliving* predecessor). This process has been shown possible in biology only as a means of fine-tuning life from previous life (a *living* DC8) under highly restricted conditions. Its operation in a cumulative manner leading to the origin of life (leading from nonlife to a living cell) remains an illusion. The universe has to "contain" the genetic information for evolution if such a step-wise process is to lead from nonlife to life. Computer simulations that have been offered in defense of this position are inadvertently deceptive and misleading.

3.35 The Meteorite Microfossil Saga

Carbonaceous meteorites are thought to be fragments of extinct comets. Extinct comets are those from which most of the volatile material has been lost after many thousands of orbits around the sun. If comets originally contained microbial life in subsurface pools and lakes, it is not unreasonable to think that one will find evidence of fossilized microorganisms in carbonaceous meteorites.

3.36 The first reports of the detection of such microbial fossils in meteorites were published in the columns of *Nature* by George Claus, Bartholomew Nagy, and others (Claus and Nagy 1961; Nagy and others 1962, 1963). As soon as these publications appeared, a vigorous campaign of refutation and denial was mounted (Fitch, Schwarcz, and Anders 1962). With a ruthless and forceful denigration of these claims on the grounds of alleged contamination of the evidence, the microfossil saga faded from view for a full 20 years. And with a veritable army of nay-sayers braying so stridently, the world became convinced that this was a scientific dead-end.

3.37 The early pioneers of microfossil discovery were thus silenced and had little choice but to recant. Wickramasinghe was told by a reliable witness that Claus was ruthlessly bullied into capitulation. Nagy also retreated somewhat while continuing to hint in his writings that it might be so, rather in the manner of Galileo Galilei's whispered "*E pursimouve*"—and yet it moves.

3.38 While such rumblings continued, the earlier work of Claus and Nagy came to be superseded in quality by investigations carried out by Hans Dieter Pflug (Pflug and Haescheke-Boyer 1979). Twenty years later, with improved techniques of sample preparation, electron microscopy, and laser ion probe spectroscopy, the signal-to-noise ratio for microfossil detection was improved by at least an order of magnitude. In 1980, Pflug corresponded with Hoyle to inform him that his own new studies corroborated Hoyle and Wickramasinghe's by now well-publicized claims of extraterrestrial microbial life. In his investigation of the Murchison meteorite, ultra-thin slices of the meteorite were placed on a perfectly clean membrane and the mineral matrix leached out with hydrofluoric acid, thus leaving any included organic structures intact. Pflug discovered a wide range of organic structures uncannily similar to terrestrial microorganisms, and with laser ion probe studies and EDAX (energy dispersive x-ray spectroscopy) analysis, he found their chemical compositions and forms to be consistent with microbial fossils. Two decades later, the assiduous studies of carbonaceous chondrites by Richard Hoover of the NASA Marshall Space Flight Center confirmed the studies of Pflug (Hoover 2011). Objections to these findings cannot now be sustained.

3.39 Microorganisms in the Stratosphere

The first experiments to recover microorganisms from the stratosphere were carried out under NASA sponsorship between 1962 and 1965. A wide range of types of microorganisms was actually recovered from the stratosphere using balloons flown to heights between 20 and 43 km. Although the lower heights in this range were not great enough to exclude terrestrial contamination, the density dependence with height of the recovered particles was consistent with the ingress of extraterrestrial microbes with an *infall* rate of 10^{19} cells per year (Greene and others 1964). This important pioneering work, carried out at the dawn of the Space Age, rang alarm bells to which the authorities had to react, and react they surely did. Wickramasinghe was told by Leslie Hale, an atmospheric scientist at Penn State University, that this exciting program of work was suddenly halted because funds were withdrawn. Nothing more was said, and no similar experiments were conducted until 2001.

3.40 The 2001–2009 ISRO-Cardiff studies of dust recovered using balloons flown to heights of 41 km in the stratosphere showed not only the presence of viable cultures of ultraviolet-resistant microorganisms but also evidence for larger 15–20 micron-sized clumps of dormant microorganisms (Wainwright and others 2003 [also Shivaji and others

2009, 2012]). The latter were detected with the use of fluorescent dyes showing the presence of viable but not culturable microorganisms. The possibility of terrestrial contamination was virtually excluded, for the height of 41 km is too high for lofting large clumps of terrestrial material from the Earth's surface. Here is evidence of a continuing incidence of microorganisms from space. Hoyle and Wickramasinghe estimated a daily input of viable biomaterial of about 0.1 ton averaged over the whole Earth.

3.41 Cometary Panspermia Today

Cometary panspermia remains a widely resisted theory. Resistance, however, has been growing progressively weaker. In *Life Everywhere: The Maverick Science of Astrobiology* (2001), David Darling contrasts a time when panspermia was ridiculed and oppressed with the present. "Today," observes Darling, "panspermia is at least tolerated in polite company" (2001: 47). This present tolerance differs remarkably from previous times. Until recently, merely mentioning the word panspermia could endanger a researcher's career. Darling notes:

> No longer does the mere mention of the word jeopardize a promising career or invite ridicule by one's scientific peers. (2001: 47)

Though skepticism prevails, the idea of microbes in outer space has entered mainstream scientific discussions.

> It's true that extreme theories, like that of Hoyle and Wickramasinghe, are still largely regarded with skepticism, but the idea of microbes being able to hop from world to world has very much entered the scientific mainstream. (Darling 2001: 47)

A theory once ridiculed and excluded is now tolerated, and even (skeptically) considered.

3.41n—According to some speculations, intolerance toward panspermia resulted in Hoyle being excluded from a share in the 1983 Nobel Prize. Along with many others, John Maddox, emeritus editor of *Nature*, found it shameful that Hoyle was awarded no share of the 1983 prize that went to Subramanyam Chandrasekhar and William A. Fowler for pioneering work in stellar nucleosynthesis (Maddox 2001, 2002). Simon Mitton writes: "Many commentators were deeply shocked when the 1983 Nobel Prize for Physics was awarded to Fowler and Subrahmanyan Chandrasekhar for contributions to nucleosynthesis and stellar structure, respectively, areas in which Hoyle's achievement exceeded that of the two laureates" (2008: 391). Recall that Hoyle co-authored with E. Margaret Burbidge, Geoffrey R. Burbidge, and William A. Fowler "the classic paper ['Synthesis of the Elements in Stars' (1957)] now known affectionately as B^2FH" (John Maddox 2001: 270 [also Martin Rees 1997: 6; Tyson and Goldsmith 2004: 165]). Chandra Wickramasinghe has written that Maddox "surmised that the reason for the exclusion was [Hoyle's] involvement in panspermia, a theory that the Swedish Academy did not wish to endorse" (2005a: 162).

3.42 Darling explains the change from strict intolerance for panspermia to reluctant tolerance. Recent discoveries of extremophiles (organism that can survive extreme conditions) make the idea of microbes surviving in space plausible. Also, consider the 1969 *Apollo 12* mission to the moon. On that mission, the astronauts recovered a lunar *Surveyor 3* camera. That instrument contained bacteria from Earth (*Streptococcus mitus*) that had survived 31 months on the lunar surface (Darling 2001: 48). Here is NASA-generated proof that microbes can survive in space.

3.42n—Concerning the significance of the *Apollo 12* moon mission, Darling quotes mission commander Pete Conrad as saying: "I always thought the most significant thing that we ever found on the whole damn Moon was that little bacteria who came back and lived and nobody ever said [an expletive] about it" (2001: 48). Saying nothing or little about "that little bacteria" remains part of NASA history. Also, see *Living Universe: NASA and the Development of Astrobiology* (2005) by Steven J. Dick (chief NASA historian) and James E. Strick.

3.43 The discovery of extremophiles, reconsiderations of the significance of bacteria recovered during the *Apollo 12* moon mission, and new evidence of rocks tossed between planets (Darling 2001: 49) render panspermia fully plausible. At minimum, Hoyle and Wickramasinghe were proven correct in insisting that microbes can survive outer space.

3.44 Though skepticism still prevails, panspermia cannot be dismissed without extended argument. As Edward Harrison remarked, "Almost certainly we have not heard the last word on the panspermia theory" (2005: 536).

3.45 More recently, in "DNA Sequencing and Predictions of the Cosmic Theory of Life" (2013), Chandra Wickramasinghe offers the following update on panspermia research:

> The theory of cometary panspermia, developed by the late Sir Fred Hoyle and the present author argues that life originated cosmically as a unique event in one of a great multitude of comets or planetary bodies

> in the Universe. Life on Earth did not originate here but was introduced by impacting comets, and its further evolution was driven by the subsequent acquisition of cosmically derived genes. Explicit predictions of this theory published in 1979–1981, stating how the acquisition of new genes drives evolution, are compared with recent developments in relation to horizontal gene transfer, and the role of retroviruses in evolution. Precisely-stated predictions of the theory of cometary panspermia are shown to have been verified.
>
> (2013: abstract)

Though cometary panspermia is not yet widely accepted, it has been strongly corroborated by a growing body of observational evidence.

– –

3.46 Concerning the Next Chapter

This third chapter has focused upon the context (modern science versus religion and theology) and the content of Hoyle and Wickramasinghe's thinking about the origins of microbial life in outer space, amid comets, stars, and interstellar dust. The next chapter will move from astrobiology and cometary biology to cosmology and theology.

3.47 Strictly speaking, cosmology concerns the *whole* universe, of which only some parts are observable to us or to any other observing parts of the cosmic whole. The effect of selecting to observe ("observation selection effect" [Bostrom 2002]) from the perspective of any part of reality is that the whole of reality must be only partly observed. Astronomy is concerned with observing some parts of reality, especially stars, planets, and galaxies. Cosmology is concerned with the whole of reality.

3.48 Though we can never observe the whole of reality, nevertheless, cosmology requires distinguishing the whole of reality from parts of reality. And cosmology requires distinguishing *all* parts from *some* part(s) and *no* part of reality. These primary mereological distinctions essential to cosmology have corresponding distinctions in theology. The next chapter concerns relations between cosmology and theology.

CHAPTER 4

Cosmology, Reality, and Panoramic Theology

4.1 Cosmic Variables

Differing conceptions of God can be distinguished and classified by reference to "cosmic variables." See Chapter VIII—"The Cosmic Variables"—in *Beyond Humanism: Essays in the Philosophy of Nature* (1975 [1937]) by Charles Hartshorne. Cosmic variables are "applicable"—in varying degrees—to "the whole scale of beings" (Hartshorne 1975 [1937]: 115, 121). Hartshorne argues that cosmic variables apply to "all members of the scale." The scale ranges "from the least particle of inorganic matter to the great universe itself" (112).

4.1n—Hartshorne argues that cosmic variables include "psychical variables" such as "sensations or feelings" and "memory" (118–21). Also see Hartshorne's *The Philosophy and Psychology of Sensation* (1934).

4.2 Variations in Reality-Inclusiveness

Reality-inclusiveness is a cosmic variable. Reality-inclusiveness is variously—that is, more or less—applicable to all that is real. Reality-inclusiveness ranges from the least inclusive quantum event to all more inclusive realities.

4.3 From among variations in reality-inclusiveness, we should distinguish whole from parts, including all parts, some parts, and no part of reality. We should therefore distinguish:

1. the all-inclusive whole of reality,
2. the set of all variously inclusive parts of reality,

3. some one or more variously inclusive part(s) of reality, and
4. no reality.

4.4 Reality-Inclusiveness and the Reality of God

These variations in reality-inclusiveness provide distinct options for conceiving the reality of God:

1. God is the all-inclusive whole of reality, the one and only comprehensive whole that transcends (includes and exceeds and is, therefore, greater than) the sum of all parts of reality.
2. God is all the parts of reality merely collectively considered, and there is no transcendent whole of reality.
3. God is some part(s) of reality (perhaps purely spiritual, or perhaps partly spiritual and partly material), and there is no transcendent whole of reality.
4. God is not real, and there is no transcendent whole of reality.

4.5 By referring to these options, we can distinguish among the following types of theism: panentheism, pantheism, classical theism, and atheism.

4.6 Panentheism

The affirmation (1) that "God is the all-inclusive whole of reality" defines *panentheism*. Accordingly, *all* parts of reality (creatures and creations) are included *in* (and "creatively synthesized" by [Hartshorne 1970]) the *divine* whole of reality (the Creator). And the divine whole is greater/better than (includes and transcends) the sum of all parts of reality.

4.7 The idea of an all-inclusive divine Creator is ancient. It is perhaps the most ancient idea in the history of philosophical theology.

4.7n—See "Ancient or Quasi-Panentheism: An Enunciation of Themes (ETCKW—God as Eternal-Temporal Consciousness, Knowing and Including the World): Ikhnaton [ca. 1375–58 B. C. E.], Hindu Scriptures [ca. 1000 B. C. E.], Lao-tse [fourth century B. C. E.?], Judeo-Christian Scriptures [fifth century B. C. E. to second century C. E.], Plato [427–347 B. C. E.]" (xiii, 29–57) in *Philosophers Speak of God* (1953) by Charles Hartshorne and William L. Reese. A panentheist idea—of the world as God's body—was present among second-century Romans. The Roman emperor and follower of Stoic philosophy, Marcus Aurelius (b. 121 CE, d. 180 CE; emperor from 161 to 180

CE), argued *for* conceiving of the universe as the body of a providential God in *The Meditations of Marcus Aurelius* (167 CE). A panentheist idea was present among fifth-century Christians. In *City of God* (412 CE), Saint Augustine argued *against* conceiving of the "World as the Body of God" (Babcock 2012: 122). Also in *Philosophers Speak of God* (1953) by Hartshorne and Reese, see "Introduction: The Standpoint of Panentheism" (1–25), "Anselm (1033–1109): God as Absolute Maximum and the Ontological Argument" (96–106), and "Epilogue: The Logic of Panentheism" (499–514).

4.8 Though the idea is ancient, the word that now labels this ancient idea is not. The word *panentheism* was constructed from the Greek (*pan* + *en* + *theos*-ism) to translate the German "*Allingottlehre*" (All-in-Gott-ism) advanced by Karl Christian Friedrich Krause (b. 1781, d. 1832).

4.8n—John W. Cooper in *Panentheism—The Other God of the Philosophers: From Plato to the Present* (2006) says: "*Panentheism* literally means 'all-in-God-ism.' This is the Greek-English translation of the German term *Allingottlehre*, 'the doctrine that all is in God.' It was coined by Karl Krause . . . The term *panentheism* did not come into common usage, however, until Charles Hartshorne popularized it in the mid-twentieth century" (26). Cooper cites *Vorlesungen über die Grundwahrheiten der Wissenshaft* ("Lectures on the Fundamental Truths of Science") (1829) and *The Ideal of Humanity and Universal Federation* (1900) by Karl Kraus (Cooper 2006: 121–22).

4.9 Today, panentheism instructed by mathematician-philosopher Alfred North Whitehead is often called "process theology." And when instructed by Whitehead *and* by Hartshorne, panentheism is called either "process theology" or "neoclassical theology."

4.10 Concerning the terms *process* and *neoclassical*, Hartshorne is a "process" theologian in the sense of being instructed by Whitehead's *Process and Reality* (1927–28). Nevertheless, Hartshorne prefers to label his Whitehead-instructed philosophy and theology as "neoclassical" rather than "process." Hartshorne takes care to avoid favoring the term "process" (becoming, change) over the equally important term "relativity" (Walker 2004: 26–27). And because "relativity" is as important as "process," Whiteheadian–Hartshornean philosophy is frequently called "process-relational philosophy" (Mesle 2008). Also, conjoining "neo" with "classical" indicates a revision of classical thinking, including classical theism. Classical theism was "too strongly influenced by Greek philosophy as medieval scholars knew that philosophy" (Hartshorne 1984b: 1). Medieval scholars appropriated and misappropriated classical Greek and classical Roman philosophies. Medieval (mis)appropriations of classical philosophies are essential to modern classical theism. For the sake of explicit contrast with classical theism,

Hartshorne calls his panentheist version of process theology "neoclassical theism" (ix).

4.10n—Also, see "neoclassical metaphysics" in Hartshorne's *The Logic of Perfection and Other Essays in Neoclassical Metaphysics* (1962).

4.11 Pantheism

The affirmation (2) that "God is all the parts of reality collectively considered" defines *pantheism*. Accordingly, all the parts of reality are not transcended by an individual whole. Pantheism affirms *no* divine *individual*. Hence, many theists and many atheists agree that pantheism is not a genuine theism.

4.12 Classical Theisms

The affirmation (3) that "God is some part(s) of reality" is the option defining *classical theisms*. Most religious theisms are among the varieties of classical theism. There can be many varieties of classical theism, including classical Greek and classical Roman polytheism.

4.13 Though many classical theists are strongly committed to monotheism, the logic of classical theism admits possible polytheism. Where God is conceived to be some part(s) of reality, logic cannot exclude conceiving of other gods. Hence, classical Greek and classical Roman theists were not being illogical in conceiving a pantheon of many gods.

4.14 Atheism

The affirmation (4) that "God is not real" is *atheism*.

4.15 Reality-Inclusiveness and Universe

Variations in reality-inclusiveness also suggest distinct options for conceiving the universe, including these:

1. The universe is the all-inclusive whole of reality that is greater than the sum of all parts.
2. The universe is all parts of reality collectively considered (= the sum of all parts).
3. The universe is some part(s) of reality (usually, relative to us, some very large part[s]).
4. The universe is not real.

4.15n—Among multiverse theorists, "universe" is (3) some part(s) of reality. In *Many Worlds in One: The Search for Other Universes* (2006), Alex Vilenkin notes that philosophers "often define the universe as 'everything there is'" (133). And he correctly remarks, "Then, of course, there cannot be any other universe" (133). Vilenkin notes, however, that physicists usually do not use the term universe in this way. For most physicists, "universe" usually refers to a "completely disjointed, self-contained" spacetime (133). "The term *multiverse* seems to have been coined by the psychologist-philosopher William James in 1895," says Mary-Jane Rubenstein in *Worlds without End: The Many Lives of the Multiverse...* (2014: 3). Though originally concerned to emphasize plurality in nature, James came to conceive that plurality is not without unity, and therefore the plural "'multiverse' still makes a 'universe'" (4, 85). And though the term multiverse was coined by James, multiverse cosmologists usually have not followed James in coming to conceive that the multiple "verses" are parts of one universe.

4.16 Panentheism Affirms One God, One Universe

Among neoclassical panentheists, the all-inclusive and transcendent whole of reality is called *God*. And the set of all parts of reality is called *universe*. The divine whole is greater than the set/sum of all parts. God includes and exceeds (transcends) the universe.

4.17 Given this understanding of the reality-inclusiveness of God and universe, we can affirm only one God and one universe. It is logically incoherent to affirm more than one presently actual all-inclusive whole of reality (Ogden 1984: 22). And it is logically incoherent to affirm more than one presently actual set of all parts of reality (Vilenkin 2006: 133).

4.18 Where the word God refers to all-inclusive reality, atheism is illogical and nonsensical. No sensation (no observation, no experience) denies what any scientific observation unavoidably confirms (at least implicitly): that we are parts of reality existing among other variously inclusive parts of reality. And according to logical/mereological analysis, all parts of reality are parts of the all-inclusive (divine) whole of reality.

CHAPTER 5

Cosmology, Panoramic Biology, and Panoramic Psychology

5.1 Cosmology: Definitions and Scope

The convergence of astronomy and biology produces astrobiology. When astrobiology extends its range of concern beyond planets and stars to the *whole* universe, astrobiology converges with cosmology.

5.2 The word *cosmology* (cosmos + logos) was constructed by mathematician-philosopher Christian Wolff (b. 1679, d. 1754). As used by Plato (in his account of Socrates in dialogue with *Timaeus*) and by many subsequent others, the word *cosmos* refers to the "universe as an ordered whole" (*Oxford English Dictionary*). Accordingly, Wolff used *cosmology* for describing his theory of the universe as an ordered whole.

5.2n—Concerning cosmology and Christian Wolff, see "Christian Wolff" by Matt Hattche (2008) and Guy Consolmagno's "Cosmic Origins" (2011).

5.3 Cosmology is clearly distinct—though not separate—from astronomy. *Cosmology* (from the Greek κοσμοσ) denotes study of the order of the whole universe. *Astronomy* (from the Latin *astrum* and Greek αστρον) denotes study of the stars. Astronomy, including study of astronomically remote galaxies and galactic structures, is within the scope of cosmology.

5.4 Indeed, cosmologists and astronomers often use the word *cosmology* to cover both cosmology and long-range astronomy. This double coverage is acknowledged by Leonard Susskind in one of his lectures. First, he acknowledges that, by strict definition, cosmology concerns "the universe as a whole" (Susskind 2009: 1:13f). Then, he adds a nonstrict definition. He says "cosmology" means the "study of things bigger than a few billion light years," and the study of things "smaller" is "astronomy" (1:13f). Hence, by strict definition, cosmology concerns the universe as a whole. By loose definition, cosmology concerns immense (relative to us) parts of the whole universe, parts "bigger than a few billion light years" (1:13f).

5.5 At larger distance scales, the universe appears to be homogeneous, smooth, and expanding. At smaller distance scales, we see inhomogeneous collapsing clumps. Therefore, it is observationally significant to distinguish smooth long-range astronomy from clumpy short-range astronomy. Additionally, it is mereologically significant to distinguish *whole* from *parts* (including parts greater than billions of light years).

5.5n—Not all observers agree on seeing a homogeneous universe. For instance, in "Cosmology: Myth or Science," 1970 Nobel Prize winner Hannes Alfvén writes: "The Universe as we see it is obviously *not* homogeneous, as the Big-Bang model requires, but consists of a multitude of galaxies" and "Hence the large-scale isotropy of the universe, which is a cornerstone in the Big-Bang cosmology, is contradicted by observations" (1977: 8). Accordingly, Alfvén concludes: "The Big-Bang conjecture is a myth, a wonderful myth maybe, which deserves a place of honor in the columbarium which already contains the Indian myth of a cyclic Universe, the Chinese cosmic egg, the Biblical myth of creation in six days, the Ptolemaic cosmological myth, and many others" (12). Furthermore, "To try to write a grand cosmic drama leads necessarily to myth" (13). Nevertheless, mythological or not, homogeneous or not, it remains important to distinguish between whole and parts.

5.6 The strict meaning of cosmology preserves the mereological distinction. Distinct from clumpy short-range astronomy and smooth long-range astronomy (loosely called cosmology), cosmology concerns the *whole* universe.

5.7 Cosmological Models

Contemporary cosmological modelling originated during the 1940s, 1950s, and 1960s. The theoretical framework was provided by the general theory of relativity (Einstein 1917). The most crucial observational foundation rested on the 1929 Hubble–Humason measurements. They measured shifts toward red in the spectral lines of distant galaxies. Hubble interpreted these shifts as the Doppler effect of galaxies receding from the observer. The Hubble law of expansion shows the recession speed V of galaxies increasing with increasing distance d:

$V = H_0 d$, where H_0 is a constant.

5.8 This important observation (Hubble's Law) required either a reversion to an Earth-centered worldview or a radical revision of our ideas of space and time. It is a matter of history now that the resolution of this dilemma hinged on Einstein's general theory of relativity (1917). Accordingly, space and time are intimately interwoven into a four-dimensional space–time continuum. And so-called matter and the curvature of space–time are interconnected.

5.9 The currently fashionable hot big bang model started with the work of Georges Edouard Lemaître (1927 [see 1.8 in this book]). He proposed an *explosion* at the beginning of the expanding universe. This explosive expansion caused the galactic red shifts observed by Hubble and Humason in 1929. More than a decade later, George Gamow worked out the details. His theory showed how the chemical elements hydrogen, deuterium, and lithium were synthesized in the earliest phases of the Universe.

5.9n—Edgar Allan Poe's 1848 proposal predates Lemaître's 1927 proposal (see 1.11 in this book). Nevertheless, from among professional scientists publishing in technical journals, Lemaître was the first to propose an exploding/expanding universe.

5.10 The work of Einstein, Lemaître, Hubble, and Gamow provides the backdrop for cosmological models developed during the 1940s, 1950s, and 1960s. In 1948, Hermann Bondi, Thomas Gold, and Fred Hoyle proposed an alternative to the big bang theory. They argued that the expanding universe observed by Hubble and Humason was actually a phenomenon that was the same from whichever point of vantage in the universe. This new cosmology was called the "steady-state" theory of the universe and remained a serious rival to the standard big bang cosmological models well into the 1960s. In order to fit the observations, however, the steady-state theory required a new physics, for it required the creation of new "matter" from an energy field as old "matter" continued to expand.

5.11 While the details of this theory (with applications to particle physics) were still being developed by Fred Hoyle and Jayant Narlikar, the supporters of big bang cosmologies were busy seeking observational criteria that might favor their theory. In 1962, a team of radio astronomers led by Martin Ryle conducted a survey of the space distribution of radio galaxies (galaxies that were strong emitters of radio waves) and found evidence for a slight increase in their space density at the greatest distances. It was a relatively minor effect, but was quickly seized upon and construed as evidence against a steady-state universe.

5.12 Although these early radio source count observations came to be contested at a later stage, a new, unexpected discovery in 1965 appeared to seal the fate of the steady-state theory of the universe. This was the serendipitous discovery of the cosmic microwave background by Penzias and Wilson (1965) at the Bell Telephone Laboratory. This background of cosmic microwave radiation, with a temperature of 2.73 degrees Kelvin, was interpreted as the "cooled off" relic radiation from the intense heat of an initial cosmic fireball. The large-scale isotropy of this background of radiation was taken to imply a uniformity of density in the early universe. Observations of minute fluctuations in the background were then interpreted as evidence of structure formation leading eventually to galaxies, clusters of galaxies, and even voids. The satellites COBE and WMAP have traced these effects in the minutest detail, all taken to be consistent with the big bang model of the universe.

5.13 A large set of observations are now interpreted to imply a big bang origin of the universe 13.7 billion years ago. Most (if not all) of the "matter" we observe with our most powerful telescopes certainly originated in a gigantic explosion. Modern cosmological data are interpreted to imply the presence of both dark matter and dark energy. The dark energy accounts for 70 percent of the total mass energy of the Universe, the dark matter for 25 percent of the mass energy, and the normal matter of galaxies account for less than 5 percent. These estimates are, however, model-dependent.

5.14 Despite the dominance of so-called concordance cosmology (cold dark matter cosmology), there are many viable alternatives. Roger Penrose's oscillating universe model is consistent with cosmic microwave background data. The Hoyle–Burbidge–Narlikar quasi-steady-state cosmology fits most of the data. And so does the hydrogravitational dynamics (HGD) cosmology of Carl Gibson and Rudy Schild.

5.14n—In the last of these models (the model now favored by Chandra Wickramasinghe), the cosmological plasma becomes unstable 0.3 million years after the big bang, when recombination from ionized to neutral gas occurs. Condensation and collapse lead to the formation of clumps of planet mass objects—10^{80} in all—a fraction of which coalesce into fast-evolving stars in which the chemical elements of life are synthesized by nuclear reactions. Such chemical elements become incorporated into the great mass of primordial planets, and it is within these objects that an origin of life would have the best chance to happen. According to HGD cosmology, dark energy is not required, and dark matter consists of life-bearing primordial planets.

5.15 All these models remain viable options. Each explains present observational data. Time will tell which model eventually prevails.

5.16 Cosmology: Where Biology and Psychology Meet Theology

Cosmology is where panoramic biology (panspermia) and panoramic psychology (panpsychism) meet panoramic theology (panentheism). Accordingly, cosmology cannot avoid theology (Murphy and Ellis 1996).

5.17 The idea of a meeting of science and theology is unacceptable to many scientists. And it is equally unacceptable to many theologians. This is partly due to our modern North Atlantic heritage. Modernity requires that we choose between two classical options (options inherited from classical Greeks and classical Romans): materialism and dualism.

5.18 Materialism reduces all that is real to physical forces and uninspired matter. Materialism requires strict adherence to atheism. Hence, for modern scientific materialists, there is no point in meeting with theologians. On the other hand, dualism recognizes a possible theistic reality, but it separates the physical world of science from the spiritual world of religion and God. Dualism allows theism only on the spiritual side of the physical–spiritual divide. Hence, modern scientific dualists and modern theological dualists agree: science concerns physics and energy; religion concerns spirits and deity; and these sharply separated realms should never meet.

5.19 We reject both classical options (materialism and dualism) offered by modernity. In constructive postmodern fashion, we affirm a neoclassical alternative Hartshorne called *panpsychism*. This neoclassical alternative has classical, preclassical, ancient, and perhaps prehistoric roots, including various non-North Atlantic roots. The term *panpsychism* emphasizes the universal/panoramic scope of psychical living influences. And because the scope of such influences is universal, its corresponding biological term is *panspermia*, a term that emphasizes the universal/panoramic scope of potential for newly emerging life.

5.20 Panspermia and Panpsychism: Dictionary Data

sperm *noun*. LME.

[Late Latin *sperma* from Greek = sperm, seed, from base of *speirein*, to sow. . . .]

panspermia *noun*. M19.

[Greek = doctrine that the elements were a mixture of all the seeds of things, from *panspermos* containing all kinds of seed, formed as PAN- + *sperma* SPERM.]

Orig., the theory that the atmosphere is full of minute germs which develop on finding a favorable environment. Now, the idea that micro-organisms or chemical precursors of life are present in outer space and able to initiate life on reaching a suitable environment.

panpsychism *noun*. L19.
[from PAN- + PSYCHISM.]

PHILOSOPHY. The doctrine or belief that all matter, however small, has a psychical aspect or component.

(*Shorter Oxford English Dictionary*, 5th edition, 2002)

5.21 The word *panspermia* comes to us from the pre-Socratic philosopher Anaxagoras (b. c. 500 BCE, d. 428 BCE). Anaxagoras was perhaps first among the Greeks to advance the idea that mind animates the physical world (Kerferd 1972; O'Leary 2008).

5.22 The word *panpsychism* was coined in the sixteenth century by Italian Renaissance philosopher Francesco Patrizi (b. 1529, d. 1597) (Skrbina 2003, 2007). In *Nova de Universis Philosophia* (1591; reprinted with changes in 1593), Francesco Patrizi advanced a Platonic philosophical alternative to the prevailing philosophies instructed by Aristotle. Patrizi employed Plato's concept of a World Soul to describe the universe as the body of the divine mind (psych) (Purnell 2004).

5.22n—In "Fancesco Patrizi" (2004), Fred Purnell says Patrizi argued that "[t]he soul of an individual living being has the same connection to its body as the World Soul has to the universe as a whole; thus the *Anima mundi* is not simply a collection of individual souls but a separate entity which vivifies the universe as a distinct reality."

5.23 Panspermia

Panspermia, meaning panoramic provisioning for the possibility of newly emerging life (biology), agrees with the theological conception of God's universal life-favoring *providence*. Provisions essential to the possibility of life are various. They include seeds, sperm, eggs, spores, cells, food, water, energy, information, and experiences.

5.24 Providential sharing is metaphysically argued for in "A Philosophy of Shared Creative Experience" (chapter 1) in *Creative Synthesis and Philosophic Method* (1970) by Charles Hartshorne, and in the "philosophy of organism" (frequently called process philosophy and process-relational philosophy) put forward in *Process and Reality: An Essay in Cosmology* (1927–28) by Alfred North Whitehead.

5.25 This *generic* panspermia (that agrees with a theology of providence and a cosmological philosophy of organism and shared creative experience) is exemplified by *cometary* panspermia (Hoyle and Wickramasinghe 1981a [1980]). *Cometary* panspermia and cosmic fine-tuning are specific factual examples of a generic provisioning for the possibility of newly emerging life.

5.26 Providence-affirming biology (generic panspermia exemplified by cometary panspermia) agrees with providence-affirming theology. The divine Creator is the universal Provider of a cosmic context in which creaturely lives are possible.

5.27 Panoramic provisioning (such as fine-tuning the cosmos), like local provisioning (such as providing seeds or education), is a creative synthetic process. Cosmic providence and local providence are "purpose driven" (Warren 2002; Templeton 1994). Our rightly providential purpose (telos) is to provide for future life and to share creative experiences (Hartshorne 1970) with human and nonhuman creatures and the divine Creator. And, purpose is psychical.

5.27n—*Evidence of Purpose: Scientists Discover the Creator* (1994), edited by John Marks Templeton, features these essays: "Dare a Scientist Believe in Design?" by Owen Gingerich, "God's Purpose in and Beyond Time" by Russell Stannard, "The Unreasonable Effectiveness of Science" by Paul Davies, "Evidence of Purpose in the Universe" by Walter R. Hearn, "Cosmology: Evidence for God or Partner for Theology?" by Robert John Russell, "Science and God the Creator" by Arthur Peacocke, "A Potent Universe" by John Polkinghorne, "The Evolution of Purpose" by John C. Eccles, "A Physiologist Looks at Purpose and Meaning in Life" by Daniel H. Osmond, and "How Blind the Watchmaker" by David Wilcox. Also, see *Science and Cosmic Purpose* (1949) by Kelvin Van Nuys; and see *Cosmic Purpose* (2014) by Toyohiko Kagawa.

5.28 Panpsychism

Panpsychism holds that all actualities have aspects or individual components that are more or less psychical/experiential. Accordingly, panpsychism has been called "psychicalism" and "universal psychicalism" (Hartshorne 1984b) and "panexperientialism" (Griffin 2007). Panexperientialism means that experience (a psychical concept) is never entirely absent from any final actuality whatsoever.

5.29 Experience is panoramic. Hence, the famous philosophical riddle—If a tree falls in the forest and no one hears it fall, does it make a sound?—is a nonstarter. Hearing is a form of eardrum-mediated feeling, a feeling/experiencing of vibrations (Goodman 2012). The fall of any tree is felt/experienced by multitudes of microscopic and cellular individuals. There are no unfelt/unexperienced events, no absolutely nonpsychical events. Hence, to claim to observe an event characterized by absolute zero experience (absolute zero psychism) is to commit the "zero fallacy" (Hartshorne 1997).

5.30 Panpsychism and Brain Functions

Evaluating psychical events can include studying brain functions. Scientists are studying connections between brain functions (involving groups of neurons) and external events. No consensus has emerged. Other unresolved questions concern interactions between observers and quantum events and the famous "collapse of the wave function."

5.31 The dual relations between "mind" and the "external world," and the "external world" and "mind," have challenged physicists, physiologists, and philosophers alike for several decades. A problem that has a firm basis in physics came into focus with the emergence of wave mechanics in the 1930s, a scientific theory that allowed the possibility of describing photons and electrons (and other elementary particles) as simultaneously both waves and particles. Moreover, the wave mechanical description of a quantum system turned out to be essentially probabilistic, giving or at least underscoring a sense of indeterminacy to the world perceived by our senses.

5.32 The permitted states and energy levels of a quantum system are defined in terms of probabilities. If an observer sets up an apparatus to observe such a system, he or she can expect to find the system in one of several permitted states—each of which is defined by a solution of a wave equation. But when such a system is *actually* observed through a *conscious* intervention, only one of these permitted states is found. The indeterminacy ceases when the observation is made. This is known as the collapse of the wave function. This phenomenon has given rise to a long-standing debate as to the causal role of the conscious observer and of consciousness itself in the "collapse."

5.32n—See *Mindful Universe: Quantum Mechanics and the Participating Observer* (2007) by Henry P. Stapp.

5.33 Consciousness in humans manifestly involves the brain and the operation of complex networks of groups of specialized cells called neurons. Basic levels of conscious or subconscious brain activity control the physiology that keeps us alive. Less tangible and less amenable to investigation are higher levels of consciousness that are responsible for our interacting "intellectually" and "deliberately" with the external world.

5.34 Pioneering studies on the workings of the human brain were carried out in the 1950s and 1960s by Nobel laureates John Eccles, A. L. Hodgkin (1964), and A. F. Huxley. Despite unravelling the architecture of the brain and finding trillions of neurons in neural circuits performing analog computations and exchanging information among body cells, scientists have not been able to resolve the deeper questions of the possibility of neurons transmitting and receiving signals from the outside world. Eccles (1992) postulates the existence of "psychons" that are associated through quantum mechanical processes with groups of neurons. One might boldly hypothesize that these *psychons* constitute the epiphenomenon on which higher conscious states depend. The brain, through such agents, may be able to connect to the external physical world and even perhaps influence the collapse of wave functions in some way.

5.34n—For a Whiteheadian–Hartshornean criticism of Eccles's (and Popper's) "inadequate understanding of psychicalism," see "Mind, Brain, and Dualism" (1981: 426) by Philip E. Devenish.

5.35 The possibility of such a connection between the conscious brains of experimenters and a physical system has been explored by groups of scientists for a while, and some controversial results have been claimed. Ibison and Jeffers (1998) showed that the outcome of a Youngian double-slit interference pattern could be altered by the experimenter's focused attention to a high degree of statistical significance. Radin and others (2012) have confirmed a similar result. Although a great deal of controversy still surrounds this work, on the face of the evidence, one cannot deny the integrity and honesty of the researchers who have been involved.

5.35n—For more about quantum events, matter, and mind, see *Quantum Mechanics and the Philosophy of Alfred North Whitehead* (2004) by Michael Epperson, and *Mind and Cosmos: Why the Materialist Neo-Darwinian Conception of Nature Is Almost Certainly False* (2012) by Thomas Nagel.

5.36 Panpsychism as Neoclassical and Constructive Postmodern

Classical panpsychism has been significantly revised and recategorized as "neoclassical" (Hartshorne 1962) and as "postmodern" (Cobb 1964).

5.37 Applying the term *neoclassical* to panpsychism indicates revising and critically appreciating classical panpsychism. And applying the term *postmodern* indicates that this critical appreciation (of a premodern concept) is influenced by critical appreciation of modern science.

5.38 A critical appreciation of modern science was offered by Alfred North Whitehead in *Science and the Modern World* (1925), *Process and Reality: An Essay in Cosmology* (1927–28), *Adventures of Ideas* (1933), and *Essays in Science and Philosophy* (1947). The term *postmodern* was used to describe Whitehead's philosophy in "From Crisis Theology to the Post-Modern World" (1964) by John Cobb Jr. Since then, for the sake of distinguishing Whiteheadian and Hartshornean thought from deconstructive postmodernism, Whitehead scholars have developed the habit of describing Whiteheadian and Hartshornean thought as "constructive postmodern" (Griffin 2007).

5.39 Panpsychism/Universal Psychicalism/ Cosmic Consciousness

Consider Chapter 2—"The Physical and the Spiritual"—in *Omnipotence and Other Theological Mistakes* (1984b) by Charles Hartshorne. Here the main question is: What is the best conception of the relations between "the physical" and "the spiritual"? (Or variously worded: What is the best conception of relations between the physical and the mental-psychological, between matter and mind, between body and soul; and thus, here is the famous body–mind/mind–body problem.) Hartshorne considers three possible answers: *dualism*, *materialism*, and *psychicalism*.

5.40 After demonstrating the failures of dualism and materialism, Hartshorne argues for *psychicalism*. Here, psychicalism means "universal psychicalism" (= panpsychism). Hartshorne agrees with physicists who dare to say "all nature is in some sense life-like." No "absolutely new principle of life" emerges from nonlife "at some point in cosmic evolution" (Hartshorne 1984b: 62). All life emerges from previous life and ultimately from the everlasting divine life.

5.41 *Materialism* reduces all inspiration and animation to inanimate matter and energy. *Dualism* restricts inspiration and animation to special parts of reality (the obviously animated animal parts, including human animals, and perhaps spirits and deities). *Psychicalism/panpsychism* universalizes inspiration and animation.

5.42 Nature of Consciousness

Consciousness associated with the human brain is a phenomenon that has come under intense scientific scrutiny in recent times. Its ultimate nature, its manifestation in the brain, and its role in the structure and content of the universe are virtually unknown.

5.43 Contemporary scientific attitudes concerning consciousness are summarized in volumes 3 and 14 of the *Journal of Cosmology*. Roger Penrose and Stuart Hameroff have developed a theory that links consciousness with the quantum space–time geometry of the universe. The emergent point of view is that consciousness is not simply an epigenetic product of evolution, but is an integral part of the physical universe. The universe itself may depend crucially upon its presence.

5.43n—See "Cosmos, Quantum Physics, and Consciousness" (2009) edited by Subhash Kak; "Consciousness and the Universe" (2011) edited by Roger Penrose and Stuart Hameroff; and see the book *Consciousness and the Universe: Quantum Physics, Evolution, Brain and Mind* (2011) by Roger Penrose, Stuart Hameroff, Henry P. Stapp, and Deepak Chopra. Also, see "Consciousness in the Universe: A Review of the 'Orch OR' Theory" (2014) by Stuart Hameroff and Roger Penrose, and in the same issue, see responses to Hameroff and Penrose by Jack A. Tuszynski (79–80) and by Deepak Chopra (81–82).

5.44 It is interesting to note that the idea of an all-pervasive cosmic consciousness is integral to many religious traditions of ancient India, particularly Buddhism. The Buddhist concept of a human being involves a merger of three distinct components: a sperm cell, an ovum, and a "packet of consciousness" derived from the universe. Similar ideas appear to be common to other religious traditions as well.

5.45 Universal Animism

Animism is a classical Latin-based term that might seem appropriate to universalizing animation. Similarly, *hylozoism* is a classical Greek-based term from the Renaissance that might seem appropriate. Hylozoism is a Neoplatonic version of the idea that all matter (*hyle*) has animation/life (*zoe*). Modern North Atlantic scholars, however, have habitually used *animism* and *hylozoism* as pejorative labels for "superstitious" primitives committing grossly pathetic fallacies. Hence, these terms are now burdened with connotations contrary to their denotative meanings.

5.46 Universal Creationism Entails Universal Evolution

Creationism is another term burdened with connotations contrary to its denotative meaning. Denotatively, creationism means that creativity (a synthetic process) is universal (Whitehead 1927–28). In *Creative Synthesis and Philosophic Method* (1970), Charles Hartshorne argues that "*To be is to create* [To be actual is to create]" (1; original italics). Universal creationism means that the categories of Creator, creatures, creations, and creativity (creative synthetic activity) are implicated in all actualities. This denotative meaning of creationism, however, is far from the popular connotative meaning.

5.46n—Concerning creativity, in a section—"The Category of the Ultimate"—of *Process and Reality: An Essay in Cosmology* (1927–28: 21–22), Alfred North Whitehead says: "'Creativity' is the universal of universals characterizing ultimate matter of fact. It is that ultimate principle by which the many, which are the universe disjunctively, become the one actual occasion, which is the universe conjunctively... Thus 'creativity' introduces novelty into the content of the many, which are the universe disjunctively. The 'creative advance' is the application of this ultimate principle of creativity to each novel situation which it originates... The ultimate metaphysical principle is the advance from disjunction to conjunction, creating a novel entity... *The many become one, and are increased by one.* In their natures, entities are disjunctively 'many' in process of passage into conjunctive unity. This Category of the Ultimate replaces Aristotle's category of 'primary substance'" (emphasis added).

5.47 In contemporary popular thought, creationism connotes antievolutionism, and evolution connotes anticreationism. It seems that, in this respect, we have not progressed beyond the English Victorians. Among Victorians, popular debate "pitted evolutionism against creationism, facts against faith" (Evans 2011). Today, throughout the North Atlantic world, evolution remains pitted against creationism.

5.48 Nevertheless, by favoring denotative meanings over popular connotative meanings, we can make progress. Evolution and creationism can be brought together in mutually affirming ways. For an example of such, see *Evolution from Space: A Theory of Cosmic Creationism* (1981b) by Hoyle and Wickramasinghe.

5.48n—Cosmic creationism is also a Confucian concept. Indeed, "cosmic creativity" is "the central theme in the *Zhongyong*" (Wan 2008: 415). See also Ames and Hall (2001) and Tu (1979, 1989).

5.49 Contrary to popular thought, evolution must be the outcome of successive generations of creatures acting creatively. "No generation can merely reproduce its ancestors" (Whitehead 1925: 188). Where creatures create, recreate, and procreate, there must be evolution. Creativity is an evolutionary process. The religious idea of the universal Creator inspiring creativity among creatures throughout creation entails universal evolution (Haisch 2010: 201). Universal creationism entails universal evolution.

5.49n—For a trinitarian theological account of creation as an evolutionary process, see "The Evolution of Creation" (Chapter VIII) in *God in Creation* (1985) by Jürgen Moltmann.

5.50 Panpsychism and Obviously Inanimate Objects

Universalizing creativity, animation (partly indeterminant response to feeling), inspiration, sentience, feeling, experience, or other mental-psychical concepts is often regarded as absurd. This is because there are many obviously inanimate objects.

5.51 Universal psychicalism/animation is saved from absurdity by observing an important distinction. We should distinguish individuals from nonindividual collections or aggregations of individuals. The individuals are more or less animated, never wholly inanimate. The nonindividual collections/aggregations of individuals are collectively inanimate.

5.52 A stick or table is a nonindividual collection of many micro-individuals. Such objects appear to be entirely inanimate only when we do not look closely. Close inspections with microscopes can reveal cellular animations. And very close inspections with more powerful instruments can reveal quantum animations and quantum decisions. Talk of quantum "indeterminancy" is the backward (antipsychical/materialist) way of acknowledging the freedom of quantum individuals to decide.

5.52n—See "Freedom as Indeterminacy" (307–9) in *Issues in Science and Religion* (1966) by Ian G. Barbour.

5.53 In his panexperientialist defense of panpsychism, David Ray Griffin notes that "such doctrines, usually under the name *panpsychism*, have been widely rejected as patently absurd" (2007: 12). Those who object to panpsychism appeal to the fact of obviously inanimate/nonpsychical objects. Griffin argues that such objections "do not apply to Whiteheadian-Hartshornean panexperientialism" (12). Their panexperientialism distinguishes between psychical individuals and nonpsychical aggregations or swarms of psychical individuals. Concerning the distinction between aggregations and individuals, Griffin writes:

> it is essential to the Whiteheadian-Hartshornean position, the more complete characterization of which is "panexperientialism with organizational duality," to distinguish between aggregational organizations, which as such have no experience or spontaneity, and "compound individuals," which do.
>
> (2007: 12)

Distinguishing aggregations from individuals is essential to a panexperientialist explanation of obviously inanimate objects.

5.54 Hartshorne explains inanimate objects by distinguishing aggregations like flocks and swarms from individual birds and bees. He begins by acknowledging that indeed tables do not feel. He remarks:

> Of course tables do not feel; but it does not follow that there is no feeling in them.
>
> (Hartshorne 1970: 142)

He explains that the feeling in unfeeling tables is like the feeling in unfeeling flocks and swarms.

> There is feeling in a flock of birds or in a swarm of bees, but the flock or the swarm feels nothing.
>
> (142)

So unfeeling tables are like unfeeling flocks and unfeeling swarms. The flock or swarm as such feels nothing. Nevertheless, feeling is present among the composite individuals.

> So there can be feeling in a swarm of molecules, though the swarm does not feel.
>
> (142)

A table is an aggregation or swarm of molecules and atoms. And though the aggregation feels nothing, this is not true of its composite individuals. Hartshorne notes that modern physics shows that "even atoms have bits of freedom" (1997: 162). Indeed, "the vibratory theory of matter banished merely inert units from science" (Hartshorne 1976: 67). There are no absolutely inanimate individuals.

5.55 Panpsychism is not absurd. All inanimate objects (which are inanimate as mere aggregates) are composed of animated individuals.

CHAPTER 6

Analogy, Metaphysics, Mythical Symbols, and Religion

6.1 Mind–Body Analogy *and* Person-to-Cell Analogy

Mind–body analogy is one of the methods employed by Hartshorne in developing his natural theology. Mind–body analogy is a premodern method.

6.2 Rāmānuja, a south Indian Brahman (traditionally b. 1017, d. 1137), employed a "body-self analogy" (Smart 1972 [1967]: 65). Experience teaches us that a human mind has a human body. Analogously, a universal mind has a universal body. Rāmānuja conceived that the universe is the body of the all-inclusive, supremely intelligent, divine self. This panentheist conception was produced by using a premodern mind–body analogy.

6.2n—In *Philosophers Speak of God* (1953), Charles Hartshorne and William L. Reese contrast the fully explicit and systematic panentheism articulated by Whitehead with approximations developed in premodern times. Medieval Indian approximations (developed by Rāmānuja, Sankara, Sri Jiva Goswami, and Krisnadas) enabled Hartshorne and Reese to conclude "that Europe scarcely came nearer until recently" (1976 [1953]: 189; also v).

6.3 Hartshorne's method of producing panentheism is distinct from any such premodern analogy. Hartshorne's mind–body analogy is supplemented with an analog from modern cellular biology. Premodern mind–body analogy plus modern cellular biology yields a constructive postmodern method of doing natural theology. This convergent method is mind–body *and* "person-to-cell analogy" (Hartshorne 1984b: 59).

6.4 The person-to-cell or "mind–nerve–cell" analogy (Hartshorne 1984b: 56) derives from modern microscope-assisted observations in biology and neuropsychology. An individual human person includes and exceeds (transcends) the sum of its cellular parts. Analogously, the all-inclusive divine individual includes and exceeds (transcends) the sum of all parts.

6.4n—Overlap between neuroscience and theology is explored by John Bowker (William Abraham 2015). See *The Sacred Neuron: The Extraordinary New Discoveries Linking Science and Religion* (2005) by John Bowker. Also, see *The Psychic Life of Micro-Organisms* (1889) by Alfred Binet.

6.5 The person-to-cell analogy comes from modern microscope-assisted biology. The mind–body analogy comes from premodern psychology and biology. Critically appreciating both premodern and modern forms of analogy is a characteristically constructive postmodern method.

6.6 From Psychology and Biology to Theology

Psychology and biology contribute to Hartshorne's theology. Each living human animal is an individual composed of many living cellular individuals. Hartshorne conceived that each "white blood corpuscle is a tiny animal" and that each nerve cell is "a single individual" (1984b: 59). An individual cell can "feel." And "feeling of feeling, the root idea of love, goes to the bottom of things" (62). Feeling goes all the way down to the least inclusive individual quantum events (Mesle 2008: 31–41). And feeling goes all the way up to the one all-inclusive individual (God).

6.7 Modern microscope-assisted biology enables us to conceive that human individuals include and exceed (transcend) cellular individuals. This modern biology-instructed idea is *analogically extended* all the way up to the all-inclusive individual, such that the all-inclusive divine individual is taken to include and exceed (transcend) all non-divine individuals. By "analogical extension" (1970: 155; also 1976), Hartshorne moves from psychology and biology to theology.

6.8 Analogy and Metaphysics

Transcendental metaphysics is about identifying logically necessary features of existence as such. Some Hartshornean metaphysicians argue (against Hartshorne) that appeals to analogy are not metaphysical appeals. For example, see "The Experience of God: Critical Reflections on Hartshorne's Theory of Analogy" (1984a) by Schubert Ogden. (See also Ogden [1975, 2008] and Devenish and Goodwin [1989: 37, note 29].) Instead of being metaphysical, appeals to analogy are said to be symbolic, hypothetical, or speculative. Without claiming (with Hartshorne) that analogical appeals can be metaphysical, we may nevertheless claim that constructive postmodern science is constantly testing the truth-value of analogies, symbols, hypotheses, and speculations.

6.8n—Concerning analogy in philosophical theology, historian of religions Charles H. Long writes: "The adequacy of analogy as a theological and philosophical tool lies at the heart of the 'de-mythologizing' problem in theology today" (1963: 15). Also, see *Analogy and Philosophical Language* (1973) by David Burrell. And see a critical discussion of "panentheistic analogy" in "Panentheism in Metaphysical and Scientific Perspective" (2004) by Philip Clayton.

6.9 Although he employs mind–body and person-to-cell analogies, Hartshorne more frequently employs transcendental metaphysical arguments. He presents such arguments in these works: *Man's Vision of God and the Logic of Theism* (1941a), *Philosophers Speak of God* (with William L. Reese, 1953), *Reality as Social Process: Studies in Metaphysics and Religion* (1953), *The Logic of Perfection and Other Essays in Neoclassical Metaphysics* (1962), *Anselm's Discovery* (1965), *Creative Synthesis and Philosophic Method* (1970), and *The Zero Fallacy and Other Essays in Neoclassical Philosophy* (1997). Hartshorne argues metaphysically from transcendental necessity to panentheism, and he argues analogically from psychology and biology to panentheism.

6.10 From Astro-Biotic and Cosmo-Biotic Reasoning to Generalized Theology

In affirming theology, albeit in a generalized and abstract form, Hoyle and Wickramasinghe appealed to biology, astrobiology, and cosmology. Consider the extreme complexity (a very specific complexity) of even the simplest life. The chances of even the simplest microbe arising from random processes are outrageously remote. The probabilities are less than 1 in 10^{1000}! The stupendous improbability of life arising from random chance argues in favor of a nonrandom cause.

6.10n—Gibson, Schild, and Wickramasinghe (2010) argued for the possibility of an abiotic origin of life on one of 10^{80} primordial planets within the first million years of the history of the big bang. Nevertheless, this is still a probabilistic miracle.

6.11 Hoyle and Wickramasinghe reasoned about the astronomical and cosmological conditions necessary for admitting the possibility of biology. Their transcendental astro-biotic and cosmo-biotic reasoning (mislabeled anthropic reasoning after 1974) brought them to a theological conclusion.

6.12 They concluded that a cosmic intelligence is necessary. The possibility of microbial life is admitted by astronomical events, including events producing heavy elements. Such astronomical events are possible only within a cosmos that is finely tuned by an interactive "all-embracing intelligence" (Hoyle 1984 [c. 1983]: 215). All-embracing intelligence (omniscience) and life-providing (providence) cosmic influence (omnipotence) are divine attributes. As advanced by Hoyle and Wickramasinghe, microbiology requires astronomy, cosmology, and theology.

6.13 In *The Relation of Biology to Astronomy* (1980b), Hoyle speculated that astronomical processes are controlled by biological concerns. He rejected the modern scientific assumption that astronomical processes signify nothing.

> Astronomers have become accustomed to thinking of the external Universe in the words of Macbeth, as being "full of sound and fury, signifying nothing."
>
> (1980b: 22)

Life-favoring astronomical and cosmological processes cannot be explained by reference to blind processes signifying nothing. Instead, we need references to intelligent control.

> Is the biological control over astronomy to be an *intelligent control* or is it to be a product of blind evolutionary processes signifying nothing? Because there are no facts one can only speculate. My personal speculation would be that the control is intelligent.
>
> (Hoyle 1980b: 23; italics added)

Hoyle's speculation—concerning biologically friendly *intelligent control* over cosmological and astronomical processes—is implicitly protheological.

6.14 Hoyle recognized that astronomical intelligence need not be cosmic/universal intelligence. In *The Intelligent Universe*, Hoyle speculated about the possibility of "intermediate intelligences" (1984 [c. 1983]: 215). There could be intermediate intelligences between human intelligences and divine intelligence. Such intermediates could be vastly greater (vastly more inclusive) than humans, yet vastly less than universal (vastly less than all-inclusive). And, according to this speculation, an intermediate intelligence (perhaps a galactic intelligence) might control some astronomical processes (Hoyle 1980b: 23). Hence, astronomical control need not imply theology. Cosmic control, however, clearly implies divine control.

6.15 Implicitly protheological astrobiology and cosmology became explicitly protheological astrobiology and cosmology in Chapter 9—"Convergence to God"—of *Evolution from Space* (1981b) by Hoyle and Wickramasinghe. Briefly put, their argument is this. Explaining the astrophysical conditions needed to admit the possibility of microbial life requires reference to an "exceedingly high" "measure of intelligence" (Hoyle and Wickramasinghe 1981b: 141–43). Such exceptional intelligence is necessary for deliberately controlling (fine-tuning) "the coupling constants of physics" (141–43). This exceedingly high intelligence includes and exceeds all others. And "like a convergent mathematical sequence of functions," this exceedingly high intelligence "has an idealized limit" (143). And this idealized limit "is *God*, and *God* is the universe" (143; original italics).

6.16 Similarly, in *Science and the Modern World* (1925), Alfred North Whitehead describes all "actuality," including divine actuality, in terms of achieving limitations or "selections from the realm of possibilities" (163; also 173–79). To be sure, for Whitehead, "God" is "the supreme ground of limitation" (179).

6.16n—For Whitehead, God is both the supreme ground of limitation and the supreme ground of creativity. Whiteheadian philosophy affirms both metaphysical principles: "the principle of limitation" and "the principle of plenitude" (Brumbaugh 1982: 10).

6.17 Hoyle and Wickramasinghe's means of generating a theological conclusion are different from Rāmānuja's amd Hartshorne's. As we have seen, Rāmānuja employed a mind–body analogy. Hartshorne employed a mind–body analogy *and* a person-to-cell analogy. And more often, Hartshorne employed transcendental metaphysics.

6.18 Hoyle and Wickramasinghe, however, made no appeal either to a mind–body analogy or to a person-to-cell analogy. And they made no appeal to transcendental metaphysics. Nevertheless, their astro-biotic and cosmo-biotic reasoning brought them to a similar theological conclusion. Like Rāmānuja, and like Hartshorne, Hoyle and Wickramasinghe concluded that the cosmos is animated by an all-embracing cosmic intelligence.

6.19 Physical Fact and Metaphysical Truth

Consider the factual requirement for an exceedingly high measure of intelligence to control/fine-tune the coupling constants (Hoyle and Wickramasinghe 1981b: 141–43). This factual requirement is fully consistent with *modal* (logically necessary) theological conclusions.

6.20 The term "modal" refers to "modal logic" (Lewis 1918) and to Hartshorne's "theory of modality" (1991 [1965]: 60–62). Modal logic is a systematic symbolic logic. It concerns relations among possibility, impossibility, contingency/factuality, and necessity. Modal-logical analysis of the conditions necessary to the possibility of any act or fact is essential to transcendental metaphysics.

6.21 The degree of fine-tuning required to control the coupling constants is so very high that it is astonishing. Nevertheless, constructive postmodern science informed by modal logic recognizes that *any* degree of tuning would be consistent with the necessary existence of God.

6.22 To be sure, any possible contingency/factuality (including any possible design) would (if actual) exemplify metaphysical necessity. According to Hartshorne, no conceivable experience could "exemplify" the "nonexistence of God." Instead, "the occurrence of anything whatever" implies divine existence (Hartshorne 1991 [1965]: 64). And reporting any factual observation implies that *we are partly inclusive parts among variously inclusive parts of the all-inclusive* (divine) *whole of reality.*

6.22n—The biblical ethical imperative (Matthew 5:43–48; 22:34–40) that we should love our neighbors and our enemies as we should love ourselves (as parts among parts of the divine whole) is founded upon metaphysical realism. Conformity to reality (not missing the mark [not sinning]) requires recognizing that neighbors, enemies, and selves are all partly inclusive parts among variously inclusive parts of the all-inclusive divine whole of reality, and that all should be valued as such.

6.23 Hoyle and Wickramasinghe show that, with astonishing factual precision (with a very high degree of fine-tuning), "the heavens [implicitly] declare" the existence of God (Psalm 19:01). To be sure, the existence of God is what "any conceivable heavens would declare"

(Hartshorne 1973 [1967]: 87). Any conceivable heavens would exemplify partly inclusive parts among variously inclusive parts of the all-inclusive divine whole of reality. Any astronomy and physics would provide contingent examples of metaphysical necessities, contingent examples of logically necessary features of existence as such.

6.23n—In "The Criterion of Metaphysical Truth and the Senses of Metaphysics" (1975), Ogden distinguishes *strict metaphysics* concerning the logically necessary features of existence as such from *metaphysics more broadly conceived* to include the necessary features of (contingent) human existence as well.

6.24 The logically necessary (modal) aspect of theology was discovered by St. Anselm. And this discovery was expressed in Anselm's *Proslogian III* (Hartshorne 1965). Almost every negative criticism of Anselm's ontological argument, however, focuses on the nonmodal argument in *Proslogian II* (Hartshorne 1965). Norman Malcolm (1960) recognizes that *Proslogian III* presents a different argument. And Hartshorne argues that the modal argument in *Proslogian III* is valid (1965 [also 1962]). Anselm's modal argument proves that, if "God" is an intelligible concept, divine existence must be necessary and divine actuality must be contingent. Unlike much of modern science, constructive postmodern science (of the radically different Whiteheadian sort) recognizes modal distinctions. Accordingly, contingent/factual data from astronomy and physics can only exemplify (not prove [demonstrate]) the logically necessary truths of metaphysics.

6.24n—See "Anselm (1033–1109): God as Absolute Maximum and the Ontological Argument" (96–106) in *Philosophers Speak of God* (1953) by Hartshorne and Reese, and Hartshorne's *Anselm's Discovery: A Re-Examination of the Ontological Proof for God's Existence* (1965) and *The Logic of Perfection and Other Essays in Neoclassical Metaphysics* (1962). Also see *Charles Hartshorne and the Existence of God* (1985) by Donald Wayne Viney, "Theology without Metaphysics?" (2008) by Schubert M. Ogden, and *Metaphysics and the Future of Theology* (2010) by William J. Meyer. (Also, Stephen Hawking observes that Kurt Gödel "attempted to formalize St Anselm's ontological proof" [2005: 1096].)

6.25 A valid metaphysical claim is exemplified and thereby confirmed by any (physical) fact. (On the other hand, a putative metaphysical claim can be shown not to be a genuine metaphysical claim by any genuinely conceivable contrary fact.) No fact or factual claim can witness against a genuine metaphysical claim. For example, consider the metaphysical claim "Something is happening." Making any observation or factual claim can only exemplify the truth of this metaphysical claim. Even thinking "Nothing is happening" provides an example of something (thinking) happening.

6.25n—Charles Hartshorne identifies "something exists" as a metaphysical truth. According to Hartshorne, "something exists" is contradicted by no conceivable experience and confirmed by "any experience whatever," and "nothing exists" is "nonsense" (confirmed by no sensation/experience) (Hartshorne 1990: 382). Furthermore, in the process of testing the hypothesis "that 'nothing exists' and 'something exists' are both logically possible propositions" (Gamwell 1990: 110), Franklin I. Gamwell demonstrates that "something exists" is not only logically possible but also logically necessary (111, 159), and that "nothing exists" is self-refuting (112). "Something exists" is a metaphysical truth in the strict sense of the term. Such a metaphysical truth is a logically necessary existential truth, a logically necessary truth about existence or reality as such.

6.26 Similarly, any conceivable physical fact would verify metaphysical panentheism. And no conceivable fact can possibly falsify the idea of an all-encompassing whole. Schubert Ogden says:

> Thus, utterances to the effect that there is a concrete whole encompassing everything else and that there are therefore concrete parts of that whole—not these parts or those, but some parts or other—could not conceivably be factually false.
>
> (2008: 152)

Any actual or conceivably actual fact would verify the idea of an all-encompassing whole. Ogden says:

> For any, even conceivable, fact that would supposedly falsify them [metaphysical theological truths], could itself only be a part of the encompassing whole, and so could never falsify but only verify them, as well as all other strictly metaphysical utterances that imply them or that they imply.
>
> (Ogden 2008: 152)

In expressing any conceivable fact, we confirm/verify—by exemplifying—a metaphysically (and analogically) founded panentheism.

6.27 For local-contextual reasons, we experience some exemplifications as more illuminating and inspiring than others. The astro-biotic, cometary, and cosmo-biotic exemplifications presented by Hoyle and Wickramasinghe are illuminating and inspiring. And so are the heavenly declarations appreciated by William Derham and by the biblical Psalmist. Derham quotes from Psalm 19:01—"The heavens declare the glory of God" (Derham 1715: 1, 5). Far more often than metaphysical arguments, these factual witnesses (exemplifications) inspire God-oriented appreciation and reverence.

6.28 Mythical Symbols

Cometary panspermia is a contemporary scientific theory. And yet, it has been expressed in ways appreciative of premodern sources, including ancient myths and symbols.

6.29 In *Cosmic Dragons: Life and Death on Our Planet* (2001), Chandra Wickramasinghe appreciates symbols from ancient Chinese mythology. In ancient China, comets and meteors were seen as fiery bringers of life and death. And they were symbolized as "cosmic dragons." These cosmic dragons were sometimes seen as harbingers of life-giving rain and abundance. At other times, cosmic dragons signaled impending disease, famine, or other catastrophe. Wickramasinghe argues that ideas about cosmic dragons should be critically appreciated. Such ideas derive from many generations of experiences with comets, meteors, and meteor showers.

6.29n—A similar conception of the potential for death or life on Earth from outer space is expressed in "Asteroid: Doomsday or Payday?" (PBS Nova television, November 20, 2013). Also, in his science fiction short story/prose poem—"The Conversation of Eiros and Charmion" (1839 [via Scott 2002: 393–97]), Edgar Allan Poe's character—"Eiros"—laments that an approaching Earth-destroying comet "which had previously inspired us with hope, was now the source of the bitterness of despair" (396).

6.30 Wickramasinghe's approach to dragon mythology is consistent with methods employed in the history of religions. For instance, consider *The Myth of the Eternal Return or, Cosmos and History* (1974) by historian of religions Mircea Eliade. Here, myths are treated as archetypal paradigms/models derived from human participation in eternally recurring natural cosmic processes.

6.30n—Also, Mircea Eliade holds that the modern "terror of history" cannot be overcome without "*faith*" in God (1974 [1949]: 160–62).

6.31 Wickramasinghe argues that ancient mythological accounts of comets and meteors are consistent with contemporary observations. Contemporary observations reveal that orbiting comets leave tube-shaped debris trails. When the orbiting Earth crosses a tube of cometary debris, we experience meteor showers.

> For instance Comet Encke gives rise to the Taurid meteor stream that is crossed every November and June, and Comet Giacobini-Zinner gives rise to the Draconids that are crossed in early October.
>
> (Wickramasinghe 2001: 7)

Comets and meteor showers are regularly featured in ancient mythology. Moreover, according to the ancients, comets and meteor showers can bring either life or death. Contemporary observations confirm this ancient witness. Comets and showers of cometary fragments do carry life-giving and life-destroying potentials.

6.32 This mythology-related rendering of cometary panspermia goes beyond mere agreement with scientific facts about comets. Wickramasinghe aims to inspire appreciation for our "deep and meaningful relationship with the universe" (2001: 5). Modern materialist prejudice—against recognizing spiritual influences—corrodes our sense of cosmic relations. And we late modern victims of light pollution suffer from chronic sky-and-star blindness. Many of us have never seen—with naked eyes—our Milky Way Galaxy. We do not sense, let alone appreciate, relationship to the larger universe.

> In days gone by, before there were conurbations and city lights, our ancestors would have been able to enjoy a deep and meaningful relationship with the universe.
>
> (Wickramasinghe 2001: 5)

Our perceptions have become blurred by light-pollution and modern intellectual prejudice.

> Nowadays our perceptions of our cosmic ancestry are blurred not only by light pollution but also by centuries of prejudice.
>
> (Wickramasinghe 2001: 5)

Light-pollution and modern intellectual prejudices inhibit appreciation of our relations to the wider universe.

> When we look up to the night sky we are lumbered with time-hallowed intellectual constructs that on the whole tend to devalue the connection that undoubtedly exists between life on Earth and the wider universe.
>
> (Wickramasinghe 2001: 5)

Relating cometary panspermia to mythology can help us recover and advance consciousness of our connections to the cosmos.

6.33 Panspermia and Theological Implications

According to cometary panspermia, water and bacteria (with genes and viruses) were transported to Earth by comets and cometary debris. And very finely tuned cosmic and stellar processes provided this life-permitting context. This is cosmic providence. Cosmic providence implies cosmic Provider.

6.34 This theological implication accounts for much modern scientific resistance to cometary panspermia.

> There is a sniff of God or Creator in the cosmic dragons and that may well be repugnant to the intelligentsia of the twenty-first century.
>
> (Wickramasinghe 2001: 183)

Our modern materialist heritage prejudices us against acknowledging theological implications.

> After a long history of industrial progress and materialism that was ruthlessly driven by reductionist Cartesian philosophy, we tend to turn away from even the vaguest hint of a creative origin of life.
>
> (Wickramasinghe 2001: 183)

This hint of God makes cometary panspermia and cosmic fine-tuning repugnant to late modern science. Even the vaguest affirmations of theology are now normally excluded from modern science.

6.35 Constructive postmodern science is different. The "radically different" (Griffin 2007) constructive postmodern science—indicated by Whitehead's *Science and the Modern World* (1925)—admits theology. Theology is especially welcome in cosmology and biology. For instance, consider Whitehead's *Process and Reality: An Essay in Cosmology* (1927–28). Here is a "cosmology" and a biology, a "philosophy of organism" (Whitehead 1927–28: xi) (now often called process philosophy or process-relational philosophy). Throughout, there is recourse to affirmative theological content. Hence, the index features two-thirds of a page of small print references to the word "God" (366). For Whitehead, cosmology and biology require theology.

6.36 Another radically different constructive postmodern scientific work is *The Liberation of Life: From the Cell to the Community* (1981) by

Charles Birch (Professor of Biology) and John B. Cobb Jr. (Professor of Theology). Here, a biologist and a theologian argue that biology entails sociology and theology.

6.37 Wickramasinghe is among the radically different. He is critically appreciative of both modern and premodern sources. He appreciates cellular biology and cosmic dragons. And he is not prejudiced against natural theology. Accordingly, his theory of cometary panspermia is a constructive postmodern scientific theory.

6.38 Panspermia and Religion

Cometary panspermia is a specific variety of panspermia. Other specific varieties include planetary panspermia, spaceship or directed panspermia, and human-directed panspermia. Planetary panspermia is Lord Kelvin's 1881 idea of seeds circulated by colliding planetary fragments. Wickramasinghe confirms the possibility of microbial transfers to Earth from the clouds of Venus (Hoyle and Wickramasinghe 1981c, 1982b; Wickramasinghe and Wickramasinghe 2012). "Directed Panspermia" holds microbes were "deliberately transmitted" to Earth by intelligent extraterrestrials (Crick and others 1973: 341). And human-directed microbial transfers to other planets are prescribed by Michael Mautner (2010) and Jacques Arnould (2010).

6.39 In contrast to specific varieties of panspermia, panspermia as such is generic, fully panoramic. In order to distinguish panspermia as such from its specific varieties, we employ a somewhat redundant phrase, generic panspermia.

6.40 If we discovered that comets and cometary debris have never reached Earth, this would falsify cometary panspermia. If we discovered that nothing can transfer from planet to planet, this would falsify planetary panspermia. Such contrafactual discoveries, however, would not falsify generic panspermia.

6.41 Generic panspermia, meaning fully panoramic provisioning for the possibility of newly emerging life, is essential to the theological idea that God is the universal Provider. Cometary panspermia is not. Nevertheless, cometary panspermia explains an ever-increasing body of observational data, and it exemplifies the generic idea of panoramic provisioning.

6.42 Because panoramic provisioning implies both cosmic and local providers, panspermia implies calls for gratitude. We owe ultimate gratitude to the one omnipresent cosmic provider (the all-inclusive Creator/Provider). And we owe lesser gratitude to the many locally present providers (the many variously less-than-all-inclusive creaturely providers). When ultimate gratitude (worship) is directed toward any less-than-all-inclusive Creator/Provider, that is a missing of the mark, a sin called idolatry (Vishio 2006).

6.43 All intelligent creatures should feel grateful. Gratitude, in response to cosmic providence (cosmic grace), is an essential religious sentiment. And gratitude and wonder (another essential religious sentiment) has been and continues to be inspired by various scientific inquiries (William James 1902 Gifford lectures; Carl Sagan 1985 Gifford lectures).

CHAPTER 7

Future Interdisciplinary Convergences with Theology: A Constructive Postmodern Trend

7.1 Astronomy, Astrobiology, Cosmology, Theology

As advanced by Hoyle and Wickramasinghe, astronomy, biology, astrobiology, and cosmology converge with theology. Microbes came to Earth from stardust and comets. And thus, microbiology is essentially dependent upon astronomy, astrochemistry, astrophysics, and study of comets. Moreover, astrobiological processes require cosmic fine-tuning (the subject matter of cosmology) by cosmic intelligence (the subject matter of theology). Reciprocally, Hartshornean natural theology (panentheism) yields biology and psychology (panspermia and panpsychism) with cosmological scope.

7.2 In addition to astronomy, astrobiology, and cosmology, there are other scientific disciplines advancing toward and converging with natural theology. Disciplines and fields advancing toward natural theology identified in part IV—"Scientific Perspectives on Natural Theology"—of *The Oxford Handbook of Natural Theology* (2013, edited by Russell Re Manning) include chemistry (John Knight), mathematics (John Polkinghorne), ecology (Christopher Southgate), psychology/mind (Fraser Watts), and sociology (Richard K. Fenn). At its most fundamental and generic level, sociology is advancing toward theology.

7.2n—We have already observed (6.36) that, in *The Liberation of Life: From the Cell to the Community* (1981), biologist Charles Birch and theologian John Cobb argue that biology entails theology and sociology. Also, relations between biology and theology are discussed in *Theological Biology* (1991) by Kenneth Cauthen, and relations between biology and sociology are discussed in *Biology as Society, Society as Biology: Metaphors* (1995), edited by Sabine Maasen, Everett Mendelshon, and Peter Wingart. In previous times, advancing cellular biology converged with chemistry and physics in works such as *The Biology of the Cell Surface* (1939) by Ernest Everett Just (Manning 1983: 322–23).

7.3 Sociology

In traditional modern science, sociology was usually restricted to the study of social relations among humans. In recent decades, however, scientists have become increasingly attentive to social relations with and among nonhuman creatures. Today, microbial societies are being recognized and studied (Ben-Jacob 2014). Nevertheless, even when sociology is not exclusively anthropological, modern sociology remains Earthbound.

7.4 In contrast to modern science, constructive postmodern science extends the scope of sociology and biology far beyond Earth. In constructive postmodern science, social studies apply at all levels. Hartshorne credits Whitehead with being first to argue for "the universality of societies in the cosmos, *at all levels*" (1950: 37; italics added; also, White 1955: 84). Recall that in *Science and the Modern World* (1925), Whitehead holds that science "is becoming the study of organisms," and biology "is the study of the larger organisms; whereas physics is the study of the smaller organisms" (103). And in *Process and Reality: An Essay in Cosmology* (1927–28), Whitehead labels his cosmology "the philosophy of organism" (18). Hence, social relations apply universally. Social relations apply all the way upward to the all-inclusive divine reality. Social relations apply at all intermediate levels, including human and cellular/microbial levels. And social relations apply all the way downward to the most elementary levels. Hartshorne notes:

> Whitehead is among the first to see that empiricism means the necessity of generalizing comparative psychology and sociology downwards to include physiology, biology, chemistry, and physics as studies of the more elementary types of sentient individuals and societies.
>
> (1950: 38)

Constructive postmodern science will continue expanding the scope of empirical sociology.

7.5 Metaphysical Sociology

By definition, metaphysical sociology has panoramic scope. For instance, see Charles Hartshorne's *Reality as Social Process: Studies in Metaphysics and Religion* (1953). Here is a social conception of reality (ontology) and a "social conception of the universe" (cosmology) (Hartshorne 1953: 29). And Hartshorne's *The Divine Relativity: A Social Conception of God* (1948) provides "a social conception of God" (theology). Metaphysical sociology converges with ontology, cosmology, and theology.

7.5**n**—Also, Anglicans and Methodists will note John Wesley's claim that "Christianity is essentially a social religion" (Collins and Vickers 2013: Sermon 24, paragraph 5).

7.6 Metaphysics of Nature

Given future interdisciplinary advances, metaphysics will be better appreciated among observational scientists, philosophers of nature, and philosophers of natural theology. In *Metaphysics and the Future of Theology: The Voice of Theology in Public Life* (2010), William J. Meyer prescribes public appeals to metaphysical aspects of theology. Future theology and future cosmology will feature more explicit metaphysics of nature.

7.7 Metaphysics of Morals

Future advances will include increasing attention to the necessity of theology in metaethics and moral theory. The metaphysics of nature and the metaphysics of morals must ultimately refer to the same all-inclusive whole of reality. The comprehensive divine reality is supremely "great" and supremely "good." God is "that than which nothing greater/better [or equally great/good] can be conceived." And reference to this "divine good" (Hartshorne 1923) is necessary for adequate moral theory. See *The Divine Good: Modern Moral Theory and the Necessity of God* (1990) by Franklin I. Gamwell. Contrary to modern moral theory, constructive postmodern moral theory recognizes the necessity of theology. As it advances, systematic moral theory will continue advancing toward and converging with natural theology.

7.7**n**—For more about theological and metaphysical necessity in moral theory, see "God and Righteousness" (1941a) by Charles Hartshorne; *Faith and Freedom* (1979) by Schubert M. Ogden; *Moral Progress: A Process Critique* *of MacIntyre* (2000) and "Future Ethics: MacIntyre and Whitehead on Moral Progress" (2004) by Lisa Bellantoni; "The Category of Freedom in the Thought of Charles Hartshorne" (2003) by Reginald David Broadnax; *Existence and the Good: Metaphysical Necessity in Morals and Politics* (2011) by Franklin I. Gamwell; and *Faith, Fallibility, and the Virtue of Anxiety* (2012) by Derek Malone-France. Also, see *Metaphysics as a Guide to Morals* (1992) by Iris Murdoch; *The Structure of Value: Foundations of Scientific Axiology* (2011 [1965]) by Robert S. Hartman; and *The Essentials of Formal Axiology* (2010) by Rem B. Edwards.

7.8 Liberty-Oriented Ethics, Cosmology, and Theology

Constructive postmodern thought recognizes that moral theory is connected to cosmology and theology. In *Process and Reality: An Essay in Cosmology* (1927–28), Alfred North Whitehead writes, "Morality of outlook is inseparably conjoined with generality of outlook" (15). Moral theory/ethics converges with cosmology and theology in *On the Moral Nature of the Universe: Theology, Cosmology, and Ethics* (1996) by theological ethicist Nancey Murphy and astronomer-cosmologist George F. R. Ellis. And convergence with cosmology is characteristic of some recent liberation-oriented theological ethics. For example, see Nancy R. Howell's *A Feminist Cosmology* (2000) and Barbara A. Holmes's *Race and the Cosmos* (2002) and *Liberation and the Cosmos* (2008). This constructive postmodern trend (liberation-oriented ethics converging with cosmology and theology) will continue.

7.8n—Also, see "Cosmic Race and Cosmic Grace: New Possibilities for Humankind" (1992) by Edwin S. Sylvest. Liberation-oriented postmodernism must reckon with the history of modernity. This history includes more than modern science. It includes modern agriculture (Smith 1991: 45–57; Henning 2005: 170), modern colonialism, modern genocide against Native Americans and others, modern racism (Tunstall 2013), and modern transatlantic slavery. Though telescope and microscope technologies mark the respective beginnings of "modern reckonings" (Derham 1715: 7–8) in astronomy and modern cellular biology, it is transatlantic slave trading that marks the beginning of modernity as such (Gilroy 1993; Walker 2004: 14–16). Transatlantic slave trading started as early as 1444. On August 8, 1444, 235 humans were purchased and subsequently shipped as cargo from Africa to Portugal (Thomas 1997: 21). Liberation-oriented ethics and theologies converge with and diverge from various modern disciplines in *Beyond the Pale: Reading Ethics from the Margins* (2011), edited by Stacey M. Floyd-Thomas and Miguel A. De La Torre. Liberation-oriented applied political science was connected to astronomy and theology by two colonial American astronomers: African-American Benjamin

Banneker (b. 1731, d. 1806) and European-American David Rittenhouse (b. 1732, d. 1796). See "The Liberating Role of Astronomy in an Old Farmer's Almanac: David Rittenhouse's 'Useful Knowledge' and a Benjamin Banneker Almanac for 1792" (Walker 2012b). For liberty's sake, astro-theology and cosmo-theology will become "astro-liberation theology" (Walker 2012b) and cosmo-liberation theology or liberation cosmology.

7.9 Narration/Storytelling

In "The Cosmic Creation Story," Brian Swimme encourages us to value "cosmic storytellers" (1988: 53). And in their narrative cosmology—-*Origins: Fourteen Billion Years of Cosmic Evolution*, Neil de Grasse Tyson and Donald Goldsmith write of telling "The Greatest Story Ever Told" (2004: 25–29). The art of truthful storytelling will continue to be appreciated in astronomy and cosmology.

7.10 Already, the Hoyle-told true story of our creation from stardust has achieved mythological status. And the true story of panspermia via comets is reclaiming its ancient mythological status. Here we are employing an understanding of myth and mythology that is contrary to the long-prevailing modern understanding.

7.11 Mythology

In *Alpha: The Myths of Creation* (1963), historian of religions Charles H. Long offers an understanding of myth that is contrary to the modern understanding that, by definition, myths are not true. Long conceives that myths are "true" stories about "reality" (11) and that "creation myths" provide "cosmic orientation" (18).

7.11n—Similarly, drawing upon historian of religions Mircea Eliade and especially upon theologian Rudolf Bultmann, theologian Schubert M. Ogden defines *myth* "by means of three closely related statements: 1. Myth is a particular way of thinking and speaking that, like other such ways, represents (i.e., *re*-presents, presents *again*) the reality presented in one basic mode of human experience. 2. The reality that myth represents is the ultimate reality presented in our original, internal, non-sensuous experience of ourselves, others, and the whole. 3. The particular way in which myth thinks and speaks of this ultimate reality is as a narrative or story determined, on the one hand, by its intention to answer the existential question of the meaning of this reality for us and, on the other hand, by its use of concepts and terms proper to the other basic mode of human experience, namely, our derived, external, sense experience of others and ourselves" (1983: 390). (Also, it is worth noticing that true stories about reality can be very different. All true narratives are created partly by historical events, partly by the narrator[s]/historian[s], and partly by perceivers [readers or hearers and viewers]. Telling the full story of the "Hundred Years' War," for instance, would require many hundreds of years. Instead of narrating every event from every perspective in a hundred years of motion pictures, the narrator selects a very few of the many truthful frames and connects these few selected frames or snapshots to tell the story. Another narrator can make a different selection of truthful frames to tell a very different true [or untrue] story.)

7.12 Hoyle and others did the science and told an essential part of the story of our creation. Their telling was "synthesis of the elements from hydrogen" (Hoyle 1946a) and "synthesis of the elements in stars" (B^2FH 1957). Twelve years later, multitudes were learning this part of our creation story from poetic musicians singing:

> We are stardust, we are golden.
> We are billion year old carbon.
> And we got to get ourselves back to the garden.
> (Crosby, Stills, Nash, & Young at Woodstock 1969)

As rendered by Crosby, Stills, Nash, & Young, this true creation story is partly art, partly science, and partly myth (i.e., a true story about creation providing cosmic orientation [Long 1963: 11–18]).

7.13 As we have seen, a mythological image is used to express cometary panspermia in Chandra Wickramasinghe's *Cosmic Dragons* (2001). Similarly, a "cosmic Uroboros" (an uroboros is a circular image of a snake, serpent, or dragon eating its tail) is used to express astronomy and cosmology in *The View from the Center of the Universe* (2006) by Joel R. Primack and Nancy Ellen Abrams. They claim that "myth is cosmology's native language" (Primack and Abrams 2006: 33). (And they cite *The Inner Reaches of Outer Space: Metaphor as Myth and as Religion* (1985) by Joseph Campbell [b. 1904, d. 1987].) And like Charles H. Long and Schubert M. Ogden, Primack and Abrams affirm that "myths *present* [re-present] reality" (2006: 34). Increasing appreciation for the representational (Ogden 1983) and orientational (Long 1963) truth-value of mythological narratives and images is a constructive postmodern scientific trend.

7.13n—A transatlantic African-American and Native American expression of the idea that life on Earth has celestial origins, coupled with hope of returning to our celestial home, is musically expressed in the religious spiritual "Swing Low, Sweet Chariot, Coming for to Carry Me Home" (written before 1862 by Wallis Willis [Wallace Willis], an African-American member of the Native

American Choctaw nation [see *Via Oklahoma: And Still the Music Flows* (2004) by Mabel Hovdahl Alexander]), and in George Clinton's expression of the same hope in his funk album *Mothership Connection* (Casablanca Records, 1975) and in other recordings (see the Mothership land in Houston, Texas on YouTube "Swing Down Houston 1976 1977 1978" [accessed July 14, 2014]). Also, concerning African and African-American postmodernism, see Walker (2004), Hogue (2013), and Monteiro-Ferreira (2014).

7.14 Graphic Images, Art, Dance, Music, Poetry, Literature

Constructive postmodern scientific trends include increasing appreciation for graphic images (Gates 2006b: 207–9; Barrow 2008; Nyíri 2013; Benson 2014). For example, consider recent developments in mathematics. Graphic images are now being used as tools for describing and solving equations. Michael Faux and S. James Gates Jr., call such images *adinkras* (2005). Adinkras have already been used to solve extremely difficult mathematical problems. Moreover, many adinkras are visually beautiful. Hence, beauty "emerges from mathematics" (Gates 2009). Mathematics and graphic arts are converging. Also, recent cosmological inquires include studying fractals, and tessellations created by graphic artist M. C. Escher (b. 1898, d. 1972). Studying mathematical graphic art is producing insight into possible cosmological structures. Mathematics, geometry, graphic arts, and cosmology are meeting and interacting. These and other interdisciplinary convergences are producing a "transformation of science" (Deloria 1979), and a "new kind of science" (Wolfram 2002), a constructive postmodern kind of science.

7.14n—According to "Native American Cosmologies" (1993) by Stephen C. McCluskey, Hopi children learn to observe and predict celestial movements and events (especially winter and summer solstice) by dancing. Hopi dance movements, along with other presentations of colors and symbols, are associated with astronomical movements and directions. Constructive postmodern education should, and perhaps will, include recovering and revising such premodern pedagogical methods. Teaching minds and teaching bodies (through dancing and other physically enacted rituals) should be convergent practices.

7.15 Additionally, advancing constructive postmodern scholars will develop greater appreciation for music, dance (Hanna 1987), poetry, and literature (E. Margaret Burbidge 1994: 1; Wickramasinghe and Ikeda 1998: 2–8; Wickramasinghe 2005: 16; Marian F. Sia 2010; Santiago Sia 2010; Begbie 2013; Bennett-Hunter 2013). Recent precursors, written by theologians instructed in literature appreciation by Katie Geneva Cannon (1995), include *Dancing with God: The Trinity from a Womanist Perspective* (2006) by Karen Baker-Fletcher; *Making a Way Out of No*

Way: A Womanist Theology (2008) by Monica A. Coleman; and *Creating Women's Theology: A Movement Engaging Process Thought* (2011), edited by Monica A. Coleman, Nancy R. Howell, and Helene Tallon Russell. And both aesthetic and ecological concerns are exhibited in *Sisters of Dust, Sisters of Spirit: Womanist Wordings on God and Creation* (1998) by Karen Baker-Fletcher; *Divine Beauty: The Aesthetics of Charles Hartshorne* (2004) by Daniel A. Dombrowski; *The Ethics of Creativity: Beauty, Morality, and Nature in a Processive Cosmos* (2005) by Brian G. Henning; and *God as Poet of the World: Exploring Process Theologies* (2008) by Roland Faber.

7.15n—The modern separation of science from art, especially the separation of science from poetry, was marked when the famous poet and natural philosopher Samuel Taylor Coleridge (b. 1772, d. 1834) argued in 1833 that his colleagues (men of science) in the British Association for the Advancement of Science were not true natural philosophers. Unlike true natural philosophers, they had failed to perceive that poetry was actualized throughout nature. See *Poetry Realized in Nature: Samuel Taylor Coleridge and Early Nineteenth-Century Science* (1981) by Trevor Harvey Levere. Coleridge's 1833 challenge inspired William Whewell's on-the-spot coining of the term "scientist" (on analogy with "artist"). See the prologue—"Inventing the Scientist"—in *The Philosophical Breakfast Club: Four Remarkable Friends Who Transformed Science and Changed the World* (2011) by Laura J. Snyder. The four remarkable friends were William Whewell, Charles Babbage, John Herschel, and Richard Jones. Also, note that Samuel Taylor Coleridge authored a famous narrative poem—"The Rime of the Ancient Mariner" (1798)—expressing theologically rooted value for all life: "*Farewell, farewell! But this I tell To thee, thou wedding-guest! He prayeth well who loveth well Both man and bird and beast. He prayeth best who loveth best, All things both great and small: For the dear God, who loveth us, He made and loveth all.*" Additionally, Coleridge authored a book on "theory of life" in which biology converged with theology—*Hints towards the Formation of a More Comprehensive Theory of Life* (1848), and a book on scientific method—*A Dissertation on the Science of Method* (1859 [1854]). Another mark of the modern separation of science from poetry and other art occurred in the year 1864 when British poet-laureate Lord Alfred Tennyson (b. 1809, d. 1892) rejected translations of Homer prepared by scientist John Herschel (b. 1792, d. 1871), thereby "signaling loudly his opinion that the scientists should leave poetry to the poets . . . one of the final bricks in the wall that came to separate art and science" (Snyder 2011: 352).

7.16 Science and Science Fiction

Science sometimes inspires science fiction. Science fiction is "the branch of fantastic literature that seeks plausibility against the background of science" (Rabkin 2012; also, Bould and Vint 2011). And science fiction sometimes inspires science and the development of new science-based technologies. For instance, the famous sci-fi writer Arthur C. Clarke originated the idea of the communication satellite (McAleer 2010). And we have already observed that Hoyle presented scientific ideas in his sci-fi novel *The Black Cloud* (1957). Indeed, Hoyle wrote 15 science fiction works (Mitton 2008: 392; Gough 2011: xiii), including *Rockets in Ursa Major: A Novel* (1959), *Comet Halley: A Novel in Two Parts* (1985), and, with John Elliot, a 1961 sci-fi television series and book—*A for Andromeda* (1962). (*A for Andromeda* became a drama/sci-fi movie in 2006.) No doubt, science will continue to inspire and be inspired by science fiction.

7.16n—Moreover, social ethics can be inspired by alternative visions of the future. Among students in Moral Theology, a required class for MDiv students at the Perkins School of Theology at Southern Methodist University, appreciation for social ethical prescriptions aimed at achieving the "abolition of poverty" (Martin Luther King Jr. 2010 [1967]: 170–75; also Luke 4:14–19) is sometimes expressed in terms of appreciation for Gene–Roddenberry-inspired visions of the future, such as Captain Picard's account of the economics of the future in the movie *Star Trek: First Contact* (1997). Science fiction contributes to the essential social ethical task of envisioning alternative futures.

7.17 Like most modern scientists, however, most modern science fiction writers fail to distinguish logically necessary existential truths (strictly metaphysical truths) from factual/contingent truths. Attention to "just the facts" (physics without explicit metaphysics) yields inadequate science and scientifically inadequate science fiction.

7.18 And like most modern scientists, science fiction writers concerned with God typically consider nothing resembling process/neoclassical theology. There is one very obvious exception, however. In her sci-fi books *Parable of the Sower* (1993) and *Parable of the Talents* (1998), Octavia E. Butler (b. 1947, d. 2006) presents an explicit change-oriented relational philosophy and theology. In *Parable of the Sower*, Butler writes: "All that you touch You Change. All that you Change Changes you. The only lasting truth is Change. God is Change" (Butler 2000 [c. 1993]: 3).

7.19 Science and Spirituality

In contrast to modern materialistic science, a constructive postmodern reenchanted science (Griffin 1988) appreciates that reality is physical and spiritual (mechanically determined and freely creative in various degrees) at all levels. Appreciating the unity of the physical and the spiritual (Hartshorne 1984b) at all levels can inspire us "*to see a world in a grain of sand, and a heaven in a wild flower*" (from William Blake's 1863 poem "Auguries of Innocence"). We require fully critical and poetic appreciation for cosmic spirituality, for our own intermediate spirituality, and for quantum spirituality. For example, see *The Universe in a Single Atom: The Convergence of Science and Spirituality* (2005) by Tenzin Gyatso, His Holiness the Fourteenth Dalai Lama.

7.19n—See appreciation for quantum, intermediate, and cosmic levels in *Atom, Man, and the Universe* (1969) by Hannes Alfvén.

7.20 Multiverse Speculations and Avoiding the G-Word

Since the beginning of the twenty-first century, speculations about the existence of infinitely many universes—collectively termed "multiverse" by Martin Rees (2001: 133, 157–81)—have become increasingly popular among cosmologists. Multiverse speculations are often employed to avoid the protheological implications of the big bang and the protheological implications of cosmic fine-tuning (Rees 2001: 161–65; Edwards 2001: xvi, 89–90, 123–39, 178, 179–202, 332; Hawking 2002: xii; Ellis2004; Vilenkin 2006: 132–35; Davies 2008 [2006]: 151; Kragh 2011; Walker 2011; Ewart 2013; Gingerich 2014: 129, 136, 142). Indeed, in *Worlds without End: The Many Lives of the Multiverse...* (2014), Mary-Jane Rubenstein begins her introductory chapter—"How to Avoid the G-Word"—by quoting cosmologist Barnard Carr: "If you don't want God, you'd better have a multiverse" (1). For many scientists, the protheological implications of cosmic fine-tuning are avoided by speculative appeals to the existence of an infinite number of universes, a multiverse in which all of the infinitely possible tunings are actualized.

7.20n—In *God and the Multiverse: Humanity's Expanding View of the Cosmos* (2014), research physicist Victor J. Stenger suggests that perhaps both of the two main alternatives to big bang cosmology (multiverse cosmology and steady-state cosmology) may have been motivated by atheism. Like many atheists who appeal to the multiverse to avoid the protheological implications of big bang cosmology, Hoyle—Stenger muses—may also have been motivated by atheism. Stenger says: "Perhaps Hoyle's atheism may have prejudiced him against the big bang and motivated him to seek another explanation" (2014: 179), and maybe Hoyle and his collaborators "objected to the implication being drawn by religious apologists, including the pope, that the big bang provided scientific evidence for divine creation" (183). Other books by Stenger include *God: The Failed Hypothesis: How Science Shows That God Does Not Exist* (2007), *The New Atheism: Taking a Stand for Science and Reason* (2009), *The Fallacy of Fine-Tuning: Why the Universe Is Not Designed for Us* (2011), *God and the Folly of Faith: The Incompatibility of Science and Religion* (2012), and *God and the Atom* (2013).

7.21 Avoidance is a characteristically late modern scientific response to protheological implications. Avoidance, however, cannot characterize Hoyle's own scientific work or attitude. When confronted with the protheological implications of his scientific researches, Hoyle—despite his atheism—considered them critically and even-handedly (Gingerich 2014: 143–44). And open-minded critical consideration moved Hoyle from atheism to theism (as we saw in 2.21–27, and 6.10–18). Critical consideration, rather than dismissive denial, is a characteristically constructive postmodern response to protheological implications.

7.21n—Avoiding cosmic fine-tuning does not avoid the metaphysical truth that any observation or experience implies the existence of God (Hartshorne 1991 [1965]: 64). See herein 6.22. Also concerning observation and God, see herein 1.6n, 3.47–48, and 9.2 (9.2 is a book review that addresses the problem of evidence for a God that is "never seen" [Sagan 2006 [1985 Gifford lecture]: 168].)

7.22 Multiverse and Implicit Generic Quasi-Steady-State

Multiverse speculations usually entail conceiving that our big bang is one of many such bangs and that our expanding universe is one of many, perhaps infinitely many, universes. A one-of-infinitely-many conception of our universe was advanced in 1965–66 by Fred Hoyle and Jayant V. Narlikar in the form of a "bubble-universe model" (Kragh 1996: 366) in "A Radical Departure from the 'Steady-State' Concept in Cosmology" and further advanced in 1993 by Fred Hoyle, Geoffrey Burbidge, and Jayant V. Narlikar in "A Quasi-Steady-State Cosmological Model with Creation of Matter." Helge Kragh notes that, in this model, matter is created by many "little big bangs" (each *little big bang* involving around 10^{16} times the mass of our sun!) (1996: 386).

7.23 Unlike the 1960s, when Hoyle and Narlikar first advanced the idea of bubble universes, today it is widely conceived that our expanding universe could be one of infinitely many inflating bubbles within a steadily bubbling cosmos termed "multiverse."

7.24 Even without reference to observations and calculations advanced by Hoyle, Burbidge, and Narlikar, any steadily creative cosmos (any steadily banging and bubbling cosmos) can be *generically* described as being in a quasi-steady-state. Temporally steady (no absolute beginning, no absolute ending, "no boundary" [Hawking 1980]) + steadily creative = quasi-steady, not absolutely steady.

7.25 Any absolutely steady cosmos would yield no bangs, no bubbles, no events, and therefore no "creative advance into novelty" (Whitehead 1978 [1927–28]: 222). Rather than being in an absolutely steady state, a steadily creative cosmos is in a quasi-steady-state.

7.26 The name branded "steady-state" and "quasi-steady-state" cosmologies—advanced by Hoyle, Burbidge, Narlikar, and others—were explicitly rejected by most cosmologists. Nevertheless, a seldom noticed generic quasi-steady-state is clearly implied in the recently popular idea of a multiverse steadily creating little big bangs and bubble universes. Unlike most, Martin Rees did notice. He noted the irony that acceptance of multiverse cosmology implicitly affirms that steady-state cosmology is not entirely wrong. In a parenthetical paragraph in *Our Cosmic Habitat*, Rees wrote:

> (Ironically, the steady statesmen may turn out to have been not 100 percent wrong, but merely too limited in envisioning the overall scale of the cosmos. I will mention in the final chapters the concept that our Big Bang could be one of many "bangs" popping off in a cosmos that on some gigantic scale persists in an eternal self-reproducing stationary state.)
>
> (2001: 83 [also 129–35, 157–81])

Hence, as Rees recognized, any multiverse cosmology with "many 'bangs' popping off in a cosmos" that "persist in an eternal self-reproducing stationary state" (2001: 83), like any other eventful no-boundary cosmology, implies a generic quasi-steady-state. Though the explicit specific brand name—"quasi-steady-state"—has been rejected, an implicit generic quasi-steady-state cosmology will endure.

7.27 Concluding Predictions

The modern separation of disciplines from other disciplines is still very much present. Nevertheless, we predict a constructive postmodern future. We predict increasing interdisciplinary advances and convergences.

7.27n—Similarly, see "Transcending the Disciplinary Boundaries" (2009) by Arran Gare in *Cosmos and History: Journal of Nature and Social Philosophy*, *The Big Questions: Tackling the Problems of Philosophy with Ideas from Mathematics, Economics, and Physics* (2009) by Steven E. Landsburg, and see moral philosophy converging with psychology in *Moral Psychology: A Contemporary Introduction* (2015) by Valerie Tiberius. Also, consider the many converging fields and disciplines (anthropology, history, religious and cultural studies, economics, biology, microbiology, epidemiology, DNA studies, agricultural studies, ecology, etc.) required to produce Charles C. Mann's *1493: Uncovering the New World Columbus Created* (2013). Moreover, an analysis of Alfred North Whitehead's account of the importance of cosmology for reality-based education is presented in *Whitehead, Process Philosophy, and Education* (1982) by Robert S. Brumbaugh, and in *Ecological Literacy: Education and the Transition to a Postmodern World* (1992) by David W. Orr.

7.28 In science, where theology is not explicit, it is merely implicit (see 4.18, 6.22). To the extent that scientists and other natural philosophers continue to value explicit over implicit, we can be fully confident in predicting continuing interdisciplinary advances toward and convergences with natural theology, especially with constructive postmodern natural theology.

> In many ways... basic religious aspirations are convergent with our own more rational endeavours as scientists to understand the Universe and the Origin of Life within it. Interestingly, the result we have now arrived at, namely the logical need for intelligence in the universe, is also consistent with the tenets of most of the major religions of the world.
>
> (Hoyle and Wickramasinghe 1990 [1988]: 143)

Appendices

Future Astrobiology

Chandra Wickramasinghe

8.1 Rigidity of Paradigms

Forecasting the future is a hazardous affair, particularly when it comes to predicting how beliefs and attitudes might change over time. Our present set of beliefs, whether they relate to morality, religion, or science, is a function of a long history that may go back decades, centuries, or even millennia.

8.2 Consider the primordial soup theory. It constitutes an adaptation of Aristotle's doctrine of spontaneous generation after that doctrine was deemed to be dead on account of the classic experiments of Louis Pasteur in 1878.

8.3 Aristotle, however, referred to experiments and observations supporting spontaneous generation. Of several pieces of evidence cited in his writings, one example goes thus:

> This (spontaneous generation) occurs in ponds, especially one near Knidos, which, so it is said, on one occasion dried up at the time of the Dog Star and all the mud was taken out; water began to collect in it as the first rains came, and at that point tiny fishes appeared, as soon as the water began. This fish was a kind of grey mullet which is not produced as the result of copulation, and its size was about that of a small sprat: none of them had either egg or milk... This evidence shows that certain fishes are produced spontaneously, and do not come out of eggs or from copulation.
>
> (Translation by A. L. Peck 1970: 13–19, 25–26)

And then there is the famous Aristotelian statement of "fireflies emerging from a mixture of warm earth and morning dew." Of course, all such early observations were either inaccurate or illusions. The primordial soup theory that demands the entire evolution from inorganic chemicals to life be confined to the ponds and oceans of planet Earth is a reincarnation of the earlier doctrine of spontaneous generation. This central dogma of biology that became firmly established by the beginning of the twentieth century was dictated by a body of evidence that has now become obsolete. New discoveries relating to early conditions on the Earth, as well as new insights into the molecular complexity of life, demands a radically new approach. But the paradigm of the primordial soup is turning out to be very difficult to overturn. As with all paradigms, once this one was established, it took on a life of its own. It is now part of a culture by which a generation of scientists has come to be recognized.

8.4 Such paradigm rigidity has a deeply rooted instinctive basis that extends across a wide spectrum of biology and may be compared to the elaborate dance patterns of bees. These highly intricate dance patterns are held from generation to generation, just like paradigms in science. As far as humans are concerned, our group behavior can be seen as a relic of our tribal ancestry. In those earlier times, a set of rituals that had no rational basis served a role in promoting the cohesion of a tribe.

8.5 Until the beginning of the seventeenth century, science clung tenaciously to the old Aristotelian idea that the Earth was the center of both the nonliving and living universe. The story of the Copernican revolution and the resistance to it that lasted for nearly two centuries is well known. While the Papacy at first resisted the Earth's demotion, it later accommodated to a post-Copernican world view, with the proviso that the living world remain firmly centered on the Earth. In subsequent ages, every challenge to this latter position came to be regarded as both antireligious and antiscientific. Extending the canvas of Genesis to embrace a larger cosmic system outside the Earth has been consistently spurned, and the old primordial soup theory has remained strictly confined to our planet. Nevertheless, the extension of Genesis to embrace a larger system, even the cosmos in its entirety, cannot be regarded as violating the spirit of Christian philosophy or being in any way a travesty of a generalized religious viewpoint. Nor can it be regarded as antiscientific so long as scientific facts are respected and accommodated within its formulation.

8.6 In the decades that followed the 1859 publication of Charles Darwin's *On the Origin of Species*, a head-on clash between science and religion began. Science came to be seen as antithetical to any form of religious beliefs that concerned the ultimate nature of the universe and of life. While the old religious views demanding special acts of creation and the fixity of species, still maintained by fundamentalists, were no longer tenable, a more generalized religious view about the nature of life could still be maintained without contradicting science. Darwin himself remained agnostic as to how the first life may have emerged from nonliving matter.

8.7 It is not necessary for Darwinian evolution to be restricted to the confines of a minuscule planet like the Earth. As we discussed in chapter 2, the Earth is an open system, and from our present understanding of astronomy, the volume of space available for the origin of life as well as for the genetic exchange that contributes to evolution is truly vast. There is even direct evidence of living cells arriving on Earth at the present time, and this evidence is likely to become consolidated and accepted in the future (Shivaji and others 2009; Wainwright and others 2003).

8.8 Clash with Religion: The Arkansas Trial

The enormous chasm between religion and science that currently exists is largely a reaction to the intransigence of fundamentalists who stubbornly insist on a literal reading of a religious text such as the Bible. No sensible scientist can possibly endorse such a view, and hard-headed fundamentalists of all religions are rightly seen as the enemies of science. Not surprisingly, when Hoyle and I published *Evolution from Space* (1981b), our book received a great deal of media attention, particularly on account of a chapter with the enigmatic heading "Convergence to God." On March 19 of that same year, the Governor of Arkansas signed into law an Act that stated: "Public schools within this State shall give balanced treatment to creation-science and to evolution-science." The U.S. Federal Government challenged the constitutional validity of this Act, and consequently there came to be a trial on this matter. A case was pending between the State of Arkansas and the Federal Government. In view of our much publicized views on the inadequacy of neo-Darwinism to explain the origin of life and evolution, the events that were now to unfold were not entirely unexpected.

8.9 In late October 1981, I received a phone call from Mr. David Williams, the State Attorney for Arkansas, who explained to me the nature of the forthcoming trial and invited me to come as an expert witness for the State. As I understood the situation, State Education Act No. 590, which required a balanced treatment of "Evolution Science" and "Creation Science," was being challenged by the American Civil

Liberties Union as infringing upon the First Amendment of the U.S. Constitution. The First Amendment, as is well known, requires a strict separation of State and Church. Although I held no brief for any particular religion or ecclesiastical group, my sympathies instantly went out to Mr. Williams, both because of my desire to defend freedom of thought and because of my dislike for the way that a limited form of neo-Darwinian evolution was being taught, as though it explained everything about the nature of life. Darwinian evolution certainly could not explain how life began from nonlife, at any rate on the Earth, and any opportunity to challenge the established position in this regard seemed welcome to me at the time.

8.10 After speaking at length to the State Attorney, I became convinced that all I was required to do in the trial was to defend the ideas that Hoyle and I had published in *Evolution from Space* (1981b). In order to be the State's expert witness, I had to rebut the claim of the American Civil Liberties Union that neo-Darwinian evolution (in its Earth-bound form) was in every respect a proven fact. I did not see an immediate reason for declining such an invitation. I had religious friends from many faiths, and I respected peoples' freedom to hold their particular beliefs, especially if such beliefs were benign. I did not think that the legitimate aspirations of religious communities of the State of Arkansas should be thwarted on scientific or pseudoscientific grounds that seemed insecure. I also believed that scientists who dismissed the possibility of a cosmic intelligence playing any creative role in the world were as bigoted as the religious fundamentalists. The case I presented essentially summarized my own scientific beliefs.

8.11 The following quotations are an extract of my testimony.

> The facts as we have them show clearly that life on Earth is derived from what appears to be an all pervasive galaxy-wide living system... Life was derived from, and continues to be driven by, sources outside the Earth, in direct contradiction to the neo-Darwinian theory that everybody is supposed to believe...
>
> It is stated according to the theory that the accumulation of copying errors, sorted out by the process of natural selection, the survival of the fittest, could account both for the rich diversity of life and for the steady upward progression from bacterium to Man... We agree that successive copying would accumulate errors, but such errors on the average would lead to a steady degradation of information... This conventional wisdom, as it is called, is similar to the proposition that the first page of Genesis copied billions upon billions of time would eventually accumulate enough copying errors and hence enough variety to produce not

merely the entire Bible but all the holdings of all the major libraries of the world... The processes of mutation and natural selection can only produce very minor effects in life as a kind of fine tuning of the whole evolutionary process....

In our view every crucial new inheritable property that appears in the course of the evolution of species must have an external cosmic origin... We cannot accept that the genes for producing great works of art or literature or music, or developing skills in higher mathematics emerged from chance mutations of monkey genes... If the Earth were sealed off from all sources of external genes: bugs could replicate till doomsday, but they would still only be bugs...

The notion of a creator placed outside the Universe poses logical difficulties, and is not one to which I can easily subscribe. My own philosophical preference is for an essentially eternal, boundless Universe, wherein a creator of life may somehow emerge in a natural way. My colleague, Sir Fred Hoyle, has also expressed a similar preference. In the present state of our knowledge about life and about the Universe, an emphatic denial of some form of creation as an explanation for the origin of life implies a blindness to fact and an arrogance that cannot be condoned.

8.12 My testimony, which was consistent with my beliefs, and wholeheartedly endorsed by Fred Hoyle, is not a source of any regret. The State of Arkansas Education Board that I was representing lost its case, however. In his summary of the judgement on January 5, 1982, Judge William R. Overton made the following statement:

In efforts to establish "evidence" in support of creation science, the defendants (The State of Arkansas) relied upon the same false premise... i.e., all evidence which criticized evolutionary theory was proof in support of creation science... While the statistical figures may be impressive evidence against the theory of chance chemical combinations as an explanation of origins, it requires a leap of faith to interpret those figures so as to support a complex doctrine which includes a sudden creation from nothing, a worldwide flood, separate ancestry of man and apes, and a young earth...

The defendants' argument would be more persuasive if, in fact, there were only two theories or ideas about the origins of life and the world.... Dr. Wickramasinghe testified at length in support of a theory that life on earth was "seeded" by comets which delivered genetic material and perhaps organisms to the earth's surface from interstellar dust far outside the solar system... While Wickramasinghe's theory about the origins of life on earth has not received general acceptance within the scientific community, he has, at least, used scientific methodology to

> produce a theory of origins which meets the essential characteristics of science.
>
> The Court is at a loss to understand why Dr. Wickramasinghe was called in behalf of the defendants. Perhaps it was because he was generally critical of the theory of evolution and the scientific community, a tactic consistent with the strategy of the defense. Unfortunately for the defense, Dr. Wickramasinghe demonstrated that the simplistic approach of the two model analysis of the origins of life is false. Furthermore, he corroborated the plaintiffs' witnesses by concluding that "no rational scientist" would believe the earth's geology could be explained by reference to a worldwide flood or that the earth was less than one million years old.

While the hard facts of science have always to be respected, the bee-dancing-type paradigms within which these facts are conceived must always be regarded as tentative and subject to change. As discussed in chapter 3 (3.25–29), the untenability of the "primordial soup" theory requires such a change, a change that will usher in a new paradigm in biology.

8.13 Emergence of Cometary Panspermia

One year before the Arkansas trial and before the publication of our book *Evolution from Space* (1981b), Fred Hoyle and I launched the theory of cometary panspermia at a 1980 conference in Maryland on "Comets and the Origin of Life" (Ponnamperuma 1981; and see Hoyle and Wickramasinghe 2000). This meeting was convened by Cyril Ponnamperuma, partly in response to a growing awareness of the ideas that were being discussed by us and partly in view of a renewed interest in the study of comets. Although our presentation was not the most welcome contribution at this meeting, a transformation of thinking in relation to the origin of life was already under way. It was coming to be recognized at long last that the primordial soup paradigm was in need of revision. At minimum, the need to include a contribution to the "soup" in the form of organic molecules from comets was reluctantly conceded. Recently, the degree of complexity as well as relevance to life of such cosmically produced molecules has stepped up to the point that the distinction between this weak form of panspermia and our original ideas of cometary panspermia is becoming blurred (Kaiser and others 2013).

8.14 The idea of cometary panspermia (the strong form of panspermia), including the transfer of fully fledged microbes and genes, continues to be resisted in orthodox scientific circles. Every new discovery in astronomy and biology since 1981, however, continues to provide unqualified support for this theory. When observations of Comet Halley were made at the time of its last perihelion in 1986, our predictions were dramatically borne out. An infrared spectrum of a desiccated bacterium agreed remarkably well with the first ever infrared spectrum of dust from a comet. The surface of the comet, examined at close quarters from the spacecraft *Giotto*, showed it to be similar in appearance to kerogen—the decay product of biology.

8.15 It is a feature of a good theory to be able to make predictions that can be verified or tested. Since 1981, a flood of new data has continued to match the predictions of the theory of cometary panspermia. This has been true for discoveries from many different disciplines—microbiology, geology, and astronomy. An incorrect theory would not be expected to have had such an uninterrupted record of successes. Sooner or later, a contradiction turns up that demands either a revocation or revision of the theory or paradigm.

8.16 Throughout the twentieth century and in the first decade of the present century, the vastness of the universe of galaxies, stars, and planets has been emphasized by every major astronomical breakthrough. In biology, the bewildering complexity of molecular arrangements in even the simplest living cell has pointed unerringly to a cosmic rather than a purely terrestrial origin of life.

8.17 It has been known for a long time that certain types of microbes possess properties that are not obviously related to the "average" conditions on Earth. Recent discoveries of "extremophiles" are now being claimed as evidence that life can indeed survive in harsh extraterrestrial environments. A temperature of 60°C was at one time thought to be an upper limit for life; now certain microbes have been discovered to survive and grow at 120°C. Bacteria can not only survive extremes of heat and cold but also withstand remarkably high doses of ionizing radiation. They are also known to exist in geothermal vents, in sulphurous hot springs, and in the dry valleys of the Antarctic. In 2010, it was discovered that cyanobacteria placed on the outside of the International Space Station survived alternations of freezing and heating, as well as exposure to harsh ionizing radiation for a full 18 months. Scarcely 40 years ago, the very existence of such microorganisms was thought utterly impossible.

8.18 Clash with Authority

The history of science has many examples of innovators whose ideas were so far ahead of their time that they failed to gain acceptance during their lifetime. Such individuals often suffered cruel penalties. Anaxagoras (500–428 BCE) famously argued that the sun was a red hot stone and that the moon was made of Earth, and for this impiety he was banished from Athens. Giordano Bruno (1548–1600), who maintained that the universe was full of inhabited planets, was condemned for heresy and burnt to death. In our more civilized world, we at least pay lip service to the libertarian principle of toleration of opinion and attempt to encourage a diversity of views. So any ostracism or obstacles we ourselves have faced in advocating panspermia must be reckoned to be mild compared to what has gone before!

8.19 Our experience over several decades has, however, brought to light a dangerous modern trend in the sociology of science. There is a tendency nowadays for authorities wielding power to withhold support or recognition of work that does not conform to orthodox opinion. If this tendency to repress unpopular ideas becomes a general pattern, new ideas will be systematically stifled by a tyrannical reign of authority (as happened throughout the Middle Ages), resulting in scientific stagnation.

8.20 Evolutionary Predictions

The predictions of the cosmic life theory that will be reviewed in this section are those related specifically to evolution, which were originally discussed by us in the period 1979–82 (Hoyle and Wickramasinghe 1978d, 1979a, 1979b, 1980, 1981a [1980], 1981b, 1981c, 1982a, 1982b). We argued then that if comets brought the first life to Earth four billion years ago, the process of microbial additions from comets must have continued throughout geological time and, consequently, played a role in evolution. Such considerations were later incorporated in a model in which genetic products of local evolution on a planet like the Earth were distributed and mixed on a galactic scale. It was argued that comet impacts, such as the one that happened at the K/T boundary 65 million years ago and that led to the extinction of the dinosaurs, cause the inevitable splash back into space of DNA fragments, fragments that carry the products of local evolution (Wallis and Wickramasinghe 2004; Napier 2004; Janaki Wickramasinghe and others 2010). Even partially destroyed DNA strands belonging to locally evolved life forms could carry the information of life far and wide (Wesson 2011; Chandra Wickramasinghe 2010). In this model, similar impact episodes and gene distribution events would happen recurrently whenever the cloud of comets surrounding our planetary system is disturbed by the gravitational effect of a passing interstellar cloud. We estimated the *average* time interval between successive impact episodes to be about 40 million years, so that from the time when life first appeared on Earth some hundred such gene distribution events would have taken place (Janaki Wickramasinghe and others 2010). We also estimated that genes from Earth would thus have infected millions of nascent planetary systems throughout the Milky Way.

8.21 Since we could not consider the Earth and our own solar system to be unique in this regard, it had to be assumed that similar gene dissemination processes operate for every life-bearing planet in the galaxy. As a consequence, the biosphere in which Darwinian evolution occurs must extend beyond our solar system to encompass a large fraction of the volume of the Milky Way. The stochastic nature of gene acquisition events resulting from encounters with molecular clouds would lead naturally to a stochastic feature of biological evolution—that is, sudden jumps—as is apparently observed in the Earth's record of life. A firm prediction was that genomes of extraterrestrial viruses and bacteria—sometimes responsible for pandemics of disease—had become incorporated in the germ lines of survivors and had provided the main driving force for biological evolution (Hoyle and Wickramasinghe 1979a, 1990). Although we were criticized at the time for advocating what appeared to be a return to a primitive superstition with regard to comets and plagues, advances in biology in the new millennium are beginning to provide striking evidence in our favor. Of particular importance is the discovery of horizontal gene transfer (HGT) as a process operating across a wide range of phyla.

8.22 Before proceeding further, I shall recap certain key predictions of the cosmic life theory with direct quotations from publications dating back to 1979, and then I shall indicate how these predictions have been verified. In *Diseases from Space* (1979a), we wrote:

> There are four steps in the complex process whereby such a virus multiplies itself the preventing of any one of which would confer immunity on an evolved plant or animal. The virus must first have an attachment protein fitted to an attachment site on the wall of a host cell. Second, the interaction of the attachment protein to the cell wall must serve eventually to strip other viral proteins away from the genetic material of the virus, which must then be afforded naked ingress to the host cell. Third, the viral genetic material must have the ability to overwrite the normal genetic program of the cell. And fourth, after multiplying in number, the new virus particles must be able to gain egress from the host cell in order to attack new cells.

> With so many opportunities to frustrate the attack of viruses, and yet with evolved life forms failing to avail themselves convincingly of these opportunities, we have—within the Earth-bound point of view—the makings of a contradiction. Immunity, such as it is, consists mostly of preventing the entry of virus particles, and then only the entry of specific viruses, not of viruses in general. It would be far more effective for host cells to develop genetically so as to prevent overwriting by viruses in general. Logically, such a process must exist, because the greater quantity of information present in the genetic material of the host must be able to overcome the combined information of many viruses. Yet such a defense is never presented, except possibly very transiently during the course of an infection, when a substance called "interferon" may have some such effect. The more likely explanation is that host cells *deliberately* avoid the apparently best form of defense. The situation is not at all that the virus is clever but that the host appears to be incorrigibly stupid. Indeed host cells even seem to invite the invasion of viruses by deliberately providing sites to which viruses can attach themselves.
>
> (Hoyle and Wickramasinghe 1979a: 153,154)

In *Proofs that Life Is Cosmic* (1982a), we wrote:

> Some commentators have claimed that pathogenic viruses cannot be incident from space, for an imagined reason which they believe overrides the many facts which prove otherwise. The argument seems on minimal thought to have the attractive quality of a one-line disproof. Viruses are specific to the cells they attack, it is said, as if to claim that human viruses are specific to human cells. While a minority of human viruses might be said to be specific to the cells of primates, most human viruses can actually be replicated in tissue cell cultures taken from a wide spectrum of animals, some indeed outside the mammals entirely. The proper statement therefore is that viruses are generally specific to the cells they attack to within about 150 million years of evolutionary history...
>
> If we had knowledge that evolution was an entirely terrestrial affair then of course it would be hard to see how viruses from outside the Earth could interact in an intimate way with terrestrially-evolved cells, but we have no such knowledge, and in the absence of knowledge all one can say is that viruses and evolution must go together. If viruses are incident from space then evolution must also be driven from space. How can this happen? Viruses do not always attack the cells they enter. Instead of taking over the genetic apparatus of the cell in order to replicate themselves, a viral particle may add itself placidly to one or other of the chromosomes. If this should happen for the sex cells of a species, mating between similarly infected individuals leads to a new genotype in their offspring, since the genes derived from the virus are copied together

> with the other genes whenever there is cell division during the growth of the offspring...
>
> A gene that happens to be useful to the adaptation of one life-form may be useless to another. Incidence from space knows nothing of such a difference, however, the gene being as likely to be added to the one form as the other. So genes that become functional in some species may exist only as nonsense genes in other species. This again is true. Genes that are useful to some species are found as redundant genes in other species. Suppose a new gene or genes to become added to the genotype (genome) of a number of members of some species. Suppose also that one or more of the genes could yield a protein or proteins that would be helpful to the adaptation of the species. The cells of those members of the species possessing the favourable new genes operate, however, in accordance with the previously existing genes, a problem arises as to how the new genes are to be switched into operation so as to become helpful to the species...As potentially favourable genes pile up more and more, a species acquires a growing potential for large advantageous change, it acquires the potential for a major evolutionary leap, thereby punctuating its otherwise continuing state of little change—its "equilibrium."
>
> (Hoyle and Wickramasinghe 1982a: 73–74)

8.23 Horizontal Gene Transfer

The cosmic theory of life *requires* that genes that are the products of evolution in some distant cosmic location (comets or planets) can, on occasion, be transferred to evolving life forms on the Earth (Hoyle and Wickramasinghe 1979a, 1982a, 1982b). In this way, evolutionary advantage or novelty can be acquired by terrestrial organisms on a stochastic basis, whenever alien genetic material carrying new information is introduced to the Earth and becomes accessible to terrestrial biology. We thus proposed an astronomical process of HGTs—transfers of genetic information across normal mating barriers on a cosmological scale—before HGT was firmly demonstrated to operate as a process within terrestrial biology.

8.24 The mechanism of Earth-bound HGT has now been amply documented (Keeling and Palmer 2008; Boto 2010). There is compelling evidence to support the once contentious view that HGT provides an important source of new genes and functions to recipient organisms and also is a driving force in evolution. It has also been recognized that the operation of HGT has foiled attempts to reconstruct ancient phylogenetic relationships in the search for a last universal common ancestor in the tree of life (Jain and others 2003). It is becoming amply clear that there was probably no such entity localized on the Earth but rather a cosmic ensemble of genes that has an antiquity comparable perhaps with the age of the universe itself (Joseph and Wickramasinghe 2011; Gibson and others 2011).

8.25 From all the available data, we can infer that sudden shifts in evolution, the emergence of new traits, and even the arrivals of new species occur through HGTs rather than by the comparatively slower neo-Darwinian process of mutations and natural selection (Keeling and Palmer 2008). Although the occurrence of neo-Darwinian evolution is not denied, it is probably dwarfed by HGTs in the longer term. The phenomenon described by biologists as "punctuated equilibrium," in which long periods of evolutionary stagnation are punctuated by sharp episodes of innovation and progress, is consistent with cosmically mediated gene transfers. The long periods of slow evolution are due to Earth-bound neo-Darwinian processes, in which no external gene inputs occurred.

8.26 The successful transfer of genetic information from one organism to another in a manner that permits transmission in a host's germ line requires a vector. The vector can take the form of a plasmid, virus, or bacterium, in which case the host and the vector are required to enter into some form of symbiosis. Eukaryotes possessing mitochondria or chloroplasts provide living evidence of HGTs that happened in the distant past, with mitochondria and chloroplasts being prokaryotic endosymbionts (Keeling and Palmer 2008). According to the evidence that will be discussed in the next section, similar symbiotic accommodation of viral genes may have occurred repeatedly in the primate ancestral line that led eventually to *Homo sapiens*.

8.27 In summation: The evidence that now exists for purely Earth-based gene transfers can be extended to transfers over a galactic scale if the biosphere within which life originates and evolves can be regarded as having cosmic dimensions. Transfers of alien genes will take place whenever the solar system (and the Earth) encounters genetic material (viruses and bacteria) from comets or planetary systems from which such material has been expelled (Wallis and Wickramasinghe 2004; Napier 2004; Janaki Wickramasinghe and others 2010). This position is, in my view, close to becoming undeniable.

8.28 Viral Sequences in Genomes

Sequencing the human genome has been one of the most outstanding scientific accomplishments of the new millennium. It has led to a wide range of discoveries that are transforming our ideas about viruses, disease, and evolution (Venter and others 2001; Gibson and others 2010). One surprise has been the discovery that the number of genes in human DNA (sequences coding for proteins) is as small as 20,000–25,000, rather than over 100,000, as had hitherto been suspected. Another surprise is that 8 percent of our DNA consists of sequences attributable to viruses, mostly endogenous retroviruses—RNA viruses that reverse-transcribe their RNA into DNA. Their significance in causing disease as well as contributing to evolution is only just coming to be understood, and many astounding correspondences with our 1979–81 statements quoted earlier cannot be overlooked.

8.29 The new evidence from genome sequence studies points to frequent episodes of retroviral infections (of which HIV is an example) not only in humans but also in almost all mammalian species. De Groot and others (2002) have identified an entire repertoire of genes known as MHC class 1 genes in chimpanzees that confer immunity against chimpanzee-derived simian immune deficiency virus. The inference is that modern chimp populations represent descendants from the survivors of a HIV-like pandemic that very nearly culled the entire ancestral chimp line in the distant past. The Hoyle–Wickramasinghe contention that HIV was an invader from space was much ridiculed when we first suggested it, but recent developments would appear to restore it at least to the realm of a reasonable hypothesis.

8.30 Following the integration of a retroviral gene sequence into a host's DNA, random mutations and the development of host immunity lead within a few generations to the cessation of infectivity. This may well be the long-term fate of the human HIV virus in the absence of any artificial medical intervention. Survivors of such major pandemics, on this picture, must carry DNA sequences of retroviral origin reflecting the history of prior infections. These viral gene sequences might then contribute to evolution in the long term, through the action of mutations and natural selection—just as we discussed in our 1981 book *Evolution from Space*. Offending viruses, once they are endogenized after episodes of infection, thus provide a storehouse of genetic information for future evolution.

8.31 The process by which viruses are endogenized and included in host genomes is not confined to retroviruses. A nonretrovial RNA transcript appears to have been incorporated in the germ line of several mammalian species, including rodents, around 40 million years ago (Horie and others 2010). Bacterial infection can also leave an imprint on genes. In a recent paper, Wang and others (2012) have shown that two immunomodulatory genes (SIGLEC) related to bacterial infection are inactive in humans but not in related primates. The conjecture is that these genes, when fully active, could have been targets for a lethal bacterial infection that nearly culled the human population in the past, perhaps 100,000 years ago.

8.32 Astronomical Spectroscopy

The correspondences of the predictions of our model with astronomical observations, discussed in chapter 2, continue to be verified as the quality of data has improved with the use of new telescopes, such as the Spitzer Space Telescope and the refurbished Hubble Space Telescope with ultradeep field-imaging capabilities. The high-resolution infrared spectra of galactic and extragalactic sources have shown spectral features of dust ("unidentified infrared bands," principally at 3.3, 6.2, 7.7, 8.2, and 11.3 micrometres) that can reasonably be interpreted only as biologically generated heteroaromatic molecules (Hoyle and Wickramasinghe 2000; Wickramasinghe 2010; Rauf and Wickramasinghe 2010). Among the most distant galaxies displaying aromatic/biomolecular infrared signatures is a high redshift infrared luminous galaxy at redshift 2.69 (Teplitz and others 2007). According to standard big bang cosmology, this galaxy emitted its light when the universe was at the tender age of two billion years. Sooner or later, such facts as these must come to be accepted and a revision of an obsolete paradigm conceded. Abiotic explanations on offer are inconsistent with the totality of astronomical findings.

8.33 Evidence of degradation products of biology also shows up in ultraviolet spectra of stars in the form of a 2175A spectral signature. Recent observations of the most distant galaxies have shown this same feature up to redshifts of 2.45, thereby demonstrating that biologically related materials were formed within 2.5 billion years of the presumed big bang origin of the universe (Elíasdóttir and others 2009; Motta and others 2002; Noterdaeme and others 2009). Very recently, the Hubble Space Telescope—enhanced with ultradeep imaging capabilities—has revealed a population of very distant galaxies at redshifts between z=8.5 and 12, implying that they formed between 700 and 400 million years after a big bang origin event within the context of the popular (concordance) cosmology (Richard Ellis and others 2012). The existence of such galaxies suggests that the advent of biological molecules may even have occurred much earlier in the history of the universe.

8.34 The presence of heavy elements synthesized in stars has not yet been confirmed in these most distant galaxies, but a high concentration of the life element carbon has been detected in a radio galaxy at redshift 5.19 (Matsuoka and others 2011). If the existence of life is judged by carbon abundance, it is possible to infer that signs of life may be found within a billion years after the big bang—ready for cometary panspermia thereafter.

8.35 Crucial Data from Meteorites and Comets

In chapter 3, we discussed the history of interpreting the possible evidence of fossilized extraterrestrial life in carbonaceous meteorites, which are the relics of comets. The first such claims of microfossils in meteorites by Claus and Nagy (1961 [also Nagy, Claus, and Hennessy 1962; Nagy, Fredriksson, Urey, Claus, Anderson and others 1963]) were quickly dismissed on the grounds of contamination, although these grounds later turned out to be false. Two decades later, H. D. Pflug's more careful studies provided very much stronger evidence of microfossils (Hoyle and Wickramasinghe 1982b), but these were also largely ignored. After the lapse of a further two decades, Richard Hoover at NASA Marshall Space Flight Center continued to discover structures in carbonaceous meteorites that he identified as fossils of cyanobacteria (Hoover and others 2005; Hoover 2011). Despite the strength of both Pflug's and Hoover's evidence, counterclaims that all these structures are most likely to be crystallographic artifacts still dominate the scientific literature.

8.36 The Polonnaruwa Meteorite

Just as in the case of the Copernican revolution of the sixteenth century, more nearly adequate scientific theories continue to reaffirm their validity, repeatedly and in many ways. Ultimately, a paradigm shift is forced into place.

8.37 Resistance to the theory of cometary panspermia has been every bit as ferocious as was the opposition to the Copernican revolution in the fifteenth and sixteenth centuries. In 1981, Hoyle and I were left in no doubt whatsoever about this. Hans D. Pflug's evidence of microbial life in meteorites constituted decisive proof of life in comets. The feeble excuses that were offered about the possibility that crystallization sometimes produced filamentary structures similar to bacterial fossils never convinced us. Nevertheless, such excuses satisfied supporters of the "primordial soup theories" to the extent that our claims for cometary panspermia continued to be ignored.

8.38 For a long time, we felt that if a meteorite contained fossilized biological structures that cannot be generated in any way other than through biological processes, then our position might at last be conceded. One biological structure that we had long attributed to comets is a class of photosynthetic microorganisms known as diatoms (Hoover and others 1986). These organisms are known to exist in prolific quantity in the lakes and oceans of the Earth, including the polar regions. They have a natural habitat in the sub-surface liquid pools in comets, and we have identified features in astronomical spectra that can be attributed to such microbes (Janaki Wickramasinghe and others 2010). There are over 100,000 distinct species of extant diatoms—marine and freshwater—each with distinctively sculptured siliceous outer coverings called frustules. These siliceous frustules are unmistakably biologically generated structures, and their existence constitutes clear and unequivocal proof of life.

8.39 It is a curious fact that diatoms appear very suddenly in the geological record some 180 million years ago. Considering the extreme stability of their silica shells, the absence of fossils in earlier epochs suggests a cometary injection, as would be expected from the theory of cometary panspermia.

8.40 In the context of the present book, serendipity cannot have a more appropriate exemplification than the discovery of fossilized diatoms in meteorites that fell over Sri Lanka—a country known to early explorers as the island of *Serendip*. This is precisely what happened. Minutes after a large fireball was seen by a large number of people in the skies over Sri Lanka on December 29, 2012, a large meteorite disintegrated and fell in the village of Araganwila, which is located a few miles away from the historic ancient city of Polonnaruwa.

8.41 At the time of entry into the Earth's atmosphere, the parent body of this Polonnaruwa meteorite would have had most of its interior porous volume filled with water, volatile organics, and possibly viable living cells. Fragments from a freshly cleaved interior surface of the Polonnaruwa meteorite were mounted on aluminum stubs and examined by Jamie Wallis under an environmental scanning electron microscope (Wallis and others 2013; N. C. Wickramasinghe and others 2013b). Images of the sample at low magnification display a wide range of structures that are distributed and enmeshed within a fine-grained matrix. They include fresh water and seawater diatoms and an extinct microbial fossil known as an acritarch. Some of these structures are deeply ingrained in the rock matrix. The range of species we found cannot be reasonably explained on the basis of contamination. Although the space origin of these rocks has been called into question by critics, studies of the distribution of oxygen isotopes leave no doubt of their cosmic origin (Wallis and others 2013).

8.42 We conclude, therefore, that the identification of fossilized diatoms in the Polonnaruwa meteorite is firmly established and unimpeachable. Since this meteorite is considered to be a cometary fragment, the idea of microbial life carried within comets and the theory of cometary panspermia is thus vindicated (Hoyle and Wickramasinghe 1981a [1980], 1981b, 1981c, 1982a, 2000; Janaki Wickramasinghe and others 2010). The universe, not humans, must have the final say in declaring what the world is really like.

8.43 The Sheffield Balloon Experiments

The existence of microbial structures in meteorites, in particular within cometary micrometeoroids that form part of a meteor stream, was recently confirmed by a team led from Sheffield, United Kingdom (Wainwright and others 2014). A balloon-borne device designed to collect impacting cometary micrometeoroids was flown to a height of 27 km in the stratosphere in June 2013 during the peak of the Perseid meteor shower. The device included an assembly of electron microscope stubs that were exposed to the stratosphere at the peak of a balloon flight for 17 minutes and thereafter securely and aseptically sealed and parachuted back to ground. The exposed stubs, once they were recovered, were examined under an electron microscope. Clear evidence was found of *infalling* microorganisms, some of which actually cratered the recipient stubs. Slow drifting of organic particles lofted from the ground was ruled out on various grounds, not least the fact that the craters on the stubs implied downward descent and impact at high speed. Biological structures, including the diatom frustules, were also discovered on the stubs. All these were falling downwards on to the stubs at high speed and must therefore have a cometary origin.

8.44 The *Rosetta* Mission to Comet 67P/Churyumov-Gerasimenko

The famous sci-fi writer and visionary "prophet of the space age" Arthur C. Clarke wrote a predictive essay—"The Twenty-First Century: A

(Very) Brief History" published in *Greetings, Carbon-based Bipeds!: Collected Essays, 1934–1998* (2000 [1999]). Concerning cometary panspermia in the year 2061, he wrote:

> *2061. The return of Halley's Comet; first landing by humans. The sensational discovery of both dormant and active life-forms vindicates Hoyle and Wickramasinghe's century old hypothesis that life is omnipresent throughout space.*
>
> (Clarke 2000: 539)

The *Rosetta* mission—and doubtless follow-up comet missions within a timeframe of decades—makes Arthur C. Clarke's 2061 date overly pessimistic. Indirect pointers to cometary microbiology are already emerging.

8.45 On November 12, 2014, the space module *Philae* landed on comet 67P/Churyumov-Gerasimenko (C-G for short). This event has been hailed as a "one big step for civilisation." The importance of this epoch-making technical achievement and its potential for unravelling our origins cannot be overstated. The scientific theory that comets are connected with the origins of life was first developed by the late Fred Hoyle and me from 1980 onwards. And we have already seen that evidence for this point of view has grown steadily over the years. Now it is widely accepted that, at the very least, the chemical building blocks of life were delivered to the Earth by comets, and this process effectively kickstarted the evolution of life on our planet. At the time of the first space mission to a comet in 1986—ESA's *Giotto* mission to Comet Halley—the prevailing point of view was that comets were lifeless inorganic snowballs. Weeks before the *Giotto* encounter on March 13, 1986, Fred Hoyle and I published a prediction that the surface of the comet would be "darker than coal," and this prediction was reported in the *London Times* of March 12, 1986. On the night of March 13, it turned out that our prediction was startlingly verified when, to the dismay of everyone, the comet did indeed turn out to be so dark as to be virtually invisible to the heavily shuttered-down cameras that had expected to photograph a bright snowfield. Comet Halley was indeed blacker than the blackest coal, and the largely organic composition of comets has come to be steadily vindicated since this time. The dark surface of comet 67P/C-G has already been established, and observations in the past months have yielded evidence consistent with biology.

8.46 An unexpectedly high rate of outgassing (H_2O) from comet 67P/C-G that was reported June 6, 2014, when the comet was at a heliocentric distance of 3.9 AU cannot be related to sublimation under solar heating, but such a rate is more readily understood in terms of the resumption of localized biological activity. Such a model was proposed by Wickramasinghe, Hoyle, and Lloyd (1996) for explaining eruptions of Comet Hale-Bopp at 6.5 AU. Even at such a great heliocentric distance, bolide impacts generate transient lakes that freeze-over. Chemoautotrophic microorganisms released into such "lakes" laden with organics will rapidly metabolize and replicate, releasing heat that might increase the initial melt volume by a factor of 10–30. Methane or carbon dioxide produced by bacteria can then build up to be eventually released through fissures in the overlying ices or at the lake edges.

8.47 With all these developments, Fred Hoyle's 1980 prognostication would appear close to being fulfilled.

> I suspect that the cosmic quality of microbiology will seem as obvious to future generations as the Sun being the centre of our solar system seems obvious to the present generation.
>
> (Hoyle 1980b: 24–25)

Critical Reviews, and an Autobiographical Astro-Theology

Theodore Walker Jr.

9.1 From William Derham's Modern *Demonstration* of God (1715) to John Wesley's Precedent for Constructive Postmodern *Exemplification* of God (1763): A Tale of Two Surveys

The following is a section from an article—"Postmodern Astro-Theology, Cometary Panspermia, and the Polonnaruwa Meteorite" in the *Journal of Cosmology*, volume 22, number 21 (March 2013). This slightly modified section is reproduced here with permission.

Modern Astro-Theology

The term "astro-theology" was used in *Astro-Theology: Or, A Demonstration of the Being and Attributes of God, from a Survey of the Heavens* (1715) by William Derham (b. 1657, d. 1735). Derham was an observational astronomer and an Anglican clergyman. He described math-and-telescope-assisted astronomy as "modern reckonings" (7–9). In Derham's person and work, modern astronomy plus theology produced modern astro-theology.

Constructive Postmodern Astro-Theology

A revision of modern astro-theology can be signaled by replacing "Demonstration" with "Exemplification" in Derham's subtitle. Accordingly, "*demonstration* of theological truths via astronomical survey" becomes "*exemplification* of theological truths via astronomical survey."

Demonstration connotes proof. According to constructive postmodern science instructed by Alfred North Whitehead's *Science and the Modern World* (1925), John B. Cobb's originating use of the term "postmodern" to refer to Whiteheadian thought (1964), and David Ray Griffin's *Whitehead's Radically Different Postmodern Philosophy* (2007), factual evidence can only *exemplify* (never prove) logically necessary existential truths (logically necessary truths about existence/reality).

Wesleyan Precedent

The constructive postmodern unwillingness to embrace factual proof/demonstration was authorized by another Anglican clergyman, John Wesley (b. 1703, d. 1791). Wesley fully appreciated modern science (Stone 2001: 156–57; Campbell 2010; Edwards 2012). He was an amateur scientist. He discovered a marine "glow worm" (Otto and Lodahl 2009: 103). He collected and studied scientific writings. He studied astronomy and cosmology (Collins 2011). He visited with scientists. He wrote about science and medicine. He did experiments. And he required that his clergy study natural philosophy (Otto and Lodahl 2009). He was seriously interested in relating natural science to theology. Wesley, however, avoided the factual demonstrations and the arguments from design (Intelligent Design arguments) that often characterized English natural theology, including Derham's modern astro-theology.

According to Wesley scholars, Wesley's avoidance of factually demonstrated theology is apparent in his selective plagiarism of writings by Derham. While lamenting Wesley's plagiarism, they appreciate a Wesleyan precedent for theological responses to natural science.

For instance, in "John Wesley's Precedent for Theological Engagement with the Natural Sciences" (2009), Randy Maddox notes that John Wesley's two-volume compendium—*A Survey of the Wisdom of God in Creation: A Compendium of Natural Philosophy* (1763)—included substantial blocks of material copied from other books, including material from Derham's *Astro-Theology: Or, A Demonstration of the Being and Attributes of God, from a Survey of the Heavens* (1715; bold added). Even though Wesley's 1763 "*Survey*" included a section from Derham's 1715 "*Survey*," Wesley excluded portions indicating that the existence of deity had been proven or demonstrated by the survey. Maddox notes that "Wesley deletes Derham's rhetorical flourish about the stupidity of those who cannot see evidence of deity in the regularity of the motion in the heavens" (2009: 22). Rather than offering a *demonstration* ("a demonstration of the being and attributes of God" [Derham 1715; also

1713]), Wesley was offering *examples* ("of the being and attributes of God").

Therefore, Wesley declined to classify his work as a proof-oriented *natural theology*. In contrast to the natural theology classification of Derham's "*Survey*" (1715), Wesley classified his Derham-enriched "*Survey*" as "*Natural Philosophy*" (1763) (Maddox 2009: 18–23). Similarly, W. Christopher Stewart notes:

> Wesley's rhetorical strategy contrasts sharply with that of Intelligent Design theorists. Wesley was careful not to overestimate the epistemic force of theistic arguments drawn from our experience of the natural world. Indeed, the whole tone of his *Survey* was to illuminate *who* God is rather than to establish *that* God is. By contrast, Intelligent Design theory inverts these priorities.
>
> (2009: 189)

> Wesley's approach to the relevance of natural science for theology is closer to what John Polkinghorne calls "*the new natural theology*," which is modest about . . . its relationship to science (abandoning all attempts to give theological answers to scientific questions) . . . and also . . . *The new natural theology* forswears all attempts to "prove" God's existence.
>
> (2009: 190; italics added)

In "Degrees of Certainty in John Wesley's Natural Philosophy" (2009), Laura Bartels Felleman compares Wesley's natural philosophy with that of his sources. Felleman's comparison reveals "an obvious and intentional removal [on Wesley's part] of condemnatory language directed at atheists" (2009: 59, also 78). In contrast with much eighteenth-century English natural theology, Wesley's natural philosophy included no attack on atheists and no claim to "demonstrative certainty" (74).

Wesley's rejection of factual *demonstrations* and his enthusiastic appreciation of factual *examples* (minus his plagiarism) provide a "precedent for theological engagement with the natural sciences" (Maddox 2009). Accordingly, this Wesleyan precedent authorizes a constructive postmodern revision of early modern astro-theology.

9.2 Carl Sagan's *The Varieties of Scientific Experience: A Personal View of the Search for God*, edited by Ann Druyan (New York: Penguin Press, 2006 [originally "The Search for Who We Are," unpublished, 1985]): A Descriptive, Evaluative, and Constructive Review

This critical review was published in the *Journal of Cosmology*, volume 20 (September 2012). With slight modifications, it is reproduced here with permission.

Carl Sagan (b. 1934, d. 1996) is well remembered for his enormously popular, Emmy and Peabody Award-winning, 13-part public television series called "Cosmos" and his bestselling book—*Cosmos* (New York: Ballantine Books, 1985). Also, many of us recall some of his other books, including his Pulitzer Prize-winning *The Dragons of Eden: Speculations on the Evolution of Human Intelligence* (New York: Random House, 1977) and *The Demon-Haunted World: Science as a Candle in the Dark* (Random House, c. 1995). Plus, we probably remember his involvement with the *Mariner*, *Viking*, *Voyager*, and *Galileo* space missions. Sagan was editor of *Icarus*, a professional journal on planetary research; cofounder and president of the Planetary Society; chairman of the Division of Planetary Sciences of the American Astronomical Society; president of the Planetology Section of the American Geophysical Union; chairman of the Astronomy Section of the American Association for the Advancement of Science; and the David Duncan Professor of Astronomy and Space Sciences and Director of the Laboratory for Planetary Studies at Cornell University.

To mark the tenth anniversary of Sagan's death, his 1985 Gifford Lectures on natural theology at Glasgow University in Scotland were published as a book by the Penguin Press in 2006. Originally, these unpublished lectures were collectively titled "The Search for Who We Are."

Sagan's Gifford Lectures were transcribed from audio tapes by his executive assistant Shirley Arden. Under the inspiration of Penguin Press editor Ann Godoff, the transcriptions were edited by Sagan's widow and two-decade collaborator Ann Druyan; then beautifully illustrated (37 illustrations, including NASA, Cassini, Hubble, and other images far superior to those available in 1985); updated with post-1985 scientific data by Illustrations Editor and Scientific Consultant Steven Soter; printed in an exceptionally well-crafted hardback book superbly designed by Amanda Dewey; and fitted with an appropriate jacket designed by Barbara de Wilde. Sagan's transcribed-edited-illustrated-updated-printed-jacketed collection now enjoys a new title—*The Varieties of Scientific Experience: A Personal View of the Search for God.*

This new title is similar to the title of the published collection of William James's 1901–2 Gifford Lectures at Edinburgh—*The Varieties of Religious Experience: A Study in Human Nature.* Indeed, Druyan offers the new title as "a tip of the hat to the illustrious tradition of the Gifford Lectures" (xv).

In the editor's introduction, Druyan reports that the "more Carl learned about nature, about the vastness of the universe and the awesome timescales of cosmic evolution, the more he was [religiously] uplifted" (ix). Druyan notes that Sagan the scientist had qualities associated with Old Testament religion, that as a child in Brooklyn he recited "the Hebrew V'Ahavta prayer from Deuteronomy at temple services: 'And you shall love the Lord your God with all your heart, with all your soul, with all your might'" (ix), and that "like some latter-day Joshua," Sagan sought to bring down walls, including "the wall of jargon that mystifies science" (ix), "the wall around our souls that keeps us from taking the revelations of science to heart" (ix), and the "wall separating science and religion" (xi). Instead of allowing religion to sustain "a protective wall around itself" (xii) by holding that "religious beliefs should be off-limits to scientific scrutiny" (xi), Sagan, says Druyan, "took the idea of God so seriously that it had to pass the most rigorous standards of [scientific] scrutiny" (x).

There are nine chapters: Chapter 1—"Nature and Wonder: A Reconnaissance of Heaven," Chapter 2—"Retreat from Copernicus: Modern Loss of Nerve," Chapter 3—"The Organic Universe," Chapter 4—"Extraterrestrial Intelligence," Chapter 5—"Extraterrestrial Folklore: Implications for the Evolution of Religion," Chapter 6—"The God Hypothesis," Chapter 7—"The Religious Experience," Chapter 8—"Crimes against Creation," and Chapter 9—"The Search." Sagan's "search for who we are" included searching for others, including terrestrial others, extraterrestrial others, intelligent extraterrestrials with radio transmitters, and even God.

Extraterrestrial Life

Though scientific searches continue to yield only inconclusive evidence of extraterrestrial life, Sagan suspected "life and intelligence are a cosmic commonplace" (195). In Chapter 3, "The Organic Universe" (where "organic" refers to the presence of complex carbon-based molecules), Sagan argued that because the universe is rich in complex carbon-based molecules prerequisite for life as we know it, explaining the origin of life requires no appeal to miraculous/divine activity. Instead, "the origin of life was in some sense easy, in some sense sitting in the laws of physics and chemistry" (99). For Sagan, the history of science shows that, "as science advances" (filling in explanatory gaps), there is "less and less" for a "God of the Gaps" to do (64). Sagan also reasoned that life is probably commonplace in an organic universe. Moreover, in "The Search for Extraterrestrial Life" in *Scientific American* (1994), Sagan said that "carbon- and water-based life-forms are the only kinds we know or can even imagine" (93), that life "seems to need liquid water, which in turn seems to require planets," and that "planetary systems are common" (99).

Extraterrestrial Intelligence

Sagan saw no evidence that there are intelligent extraterrestrials (evidence from UFO reports and theories about ancient astronauts were rejected as "folklore"), but he found the idea of such plausible, and he advocated scientific searches for electromagnetic signals from intelligent extraterrestrials. Sagan noted that this mainstream scientific approach was suggested by the Drake equation—N (number of technological civilizations in our galaxy capable of interstellar contact) = R (galactic star formation rate) × f_p (stars with planets) × n_p (number of planets) × f_l (planets with life) × f_i (planets with intelligent life) × f_c (civilization with communications technology) × L (lifetimes of technical civilizations) (109). According to the Drake equation, the product of various individual probabilities yields a collective probability of a number—-N—of technological extraterrestrial civilizations capable of interstellar communications. Depending upon the probabilities, there may be up to a million such civilizations in our galaxy (114–15).

Because the idea of such technological extraterrestrial civilizations implies equal-to-human or more probably superior-to-human intelligence, Sagan saw favoring this idea as a Copernicus-like threat to valuing humans as central, privileged, and superior. Sagan argued that the history of science shows that we tend to retreat from Copernican insights, from Darwinian insights, and from other scientific insights (including special relativity and big bang cosmology) that deny us cosmic centrality and privilege. Recent manifestations of this tendency include creation science, arguments from design and anthropic principles, and some arguments against the possibility of extraterrestrial intelligence. Instead of continually retreating from Copernicus-like threats to human centrality, Sagan prescribed that we adopt a "cosmic perspective" (30). Druyan says he invited us to "embrace our lack of centrality" and "face the vastness" (xii).

God and Extraterrestrials

Sagan scrutinized the cosmological argument, the arguments from the second law of thermodynamics, the argument from design, the moral argument, the ontological argument, the argument from consciousness, and the arguments from religious experiences (154–63), and he concluded that these "alleged natural theological arguments" are "not very compelling" (165).

Moreover, Sagan held that, except for slight "details," the question of "convincing evidence" for the existence of God is "not significantly different" from the question of convincing evidence for the existence of intelligent extraterrestrials (108). Moreover, "a reasonably competent god" could have provided "absolutely clear-cut evidence of His existence" (165). For example, "God could have engraved the Ten Commandments on the Moon," or God could have placed "a hundred-kilometer crucifix in Earth orbit" (167). The lack of such clearly visible evidence brought Sagan to conclude that Protagoras was correct when (in his fifth-century BCE *Essay on the Gods*) he wrote that he had "no means of knowing" if the gods exist because "they are never seen" (168). Though Sagan imagined that someday soon we may discover convincing evidence for the existence of intelligent extraterrestrials (as he argues [with co-author I. S. Shklovskiĭ] in *Intelligent Life in the Universe* [1966] and imagines in *Contact: A Novel* [1985], a work [now a motion picture] about receiving the first radio signals from extraterrestrials), Sagan did not imagine there would ever be convincing evidence for the existence of God.

"What We Mean by God"

Sagan noted that the outcome of scientific inquiry into the existence of God hinges upon "what we mean by God" and that the word "'god' is used to cover a vast multitude of mutually exclusive ideas" (224). According to Sagan, ideas of God range from, at one pole, "the view of, say, Spinoza or Einstein, which is more or less God as the sum total of the laws of physics" (and "If that's what we mean by God, then surely God exists") to the "opposite pole" where God is imaged as a man with "a long white beard, sitting in a throne in the sky" (and "for that kind of god I maintain there is no evidence") (224).

Sagan rejected the idea of "an exhortatory god" (191) or "a personal god" who gives us purpose because "purpose is not imposed from the outside" and not found "in some book written thousands of years ago" (227). Sagan held that we create our purposes, and "survival" is one purpose "that we have to work out for ourselves" (227). The scientific evidence shows that survival is far from guaranteed. We are now fully able and increasingly likely to destroy ourselves with nuclear war or by destroying our environment. Sagan drew upon religious language to describe accumulating weapons of mass destruction and increasing ecological destructions as "crimes against creation" (the title of Chapter 8). And he prescribed that religious teachings about creation, stewardship, hope, and love (including love of enemies) should be employed to assist scientists in encouraging peace and ecological responsibility (205–9). (Similarly, in *Billions and Billions: Thoughts on Life and Death at the Brink of the Millennium* [1997], Sagan prescribed a science–religion alliance aimed at protecting the Earth [172].) Sagan rejected the idea that God (or extraterrestrials) will determine human destiny because this idea discourages us from taking immediate and full responsibility for our future survival (205–9 [also 59, 129]).

Deity as "Too Small"

As an astronomer, Sagan adopted a to-scale perspective on our size, age, and noncentrality relative to our solar system, our Milky Way galaxy, and the universe. From this "cosmic perspective," Sagan found that "the God portrayed" in "Earth-centered" theology is "too small"—"a god of a tiny world and not a god of a galaxy, much less of a universe" (30). Sagan judged that portraying God as "too small" is "a general problem with much of Western theology," a problem "that theologians have not adequately addressed" (30).

Sagan rightly identified "smallness" (conceiving of a deity "too small" for universal influence) as a "general problem" for classical Western theology (30). Despite having identified "smallness" as the problem, Sagan failed to explore the implied solution—"largeness" (conceiving of a God *not* "too small" for universal influence). Sagan failed to consider the neoclassical meaning of "God is great," namely, "God is large" beyond possible equality with or surpassing by any other(s). God is large enough to encompass all that is real, however large that may be. In contrast to the Earth-centered classical theology scrutinized by Sagan, the neoclassical theology of Charles Hartshorne holds that God is "the one universal individual" (1953: 176). Schubert M. Ogden's formulation of this panentheist-Hartshornean-neoclassical doctrine is that God is "the one all-inclusive whole of reality" (1984: 21). According to neoclassical theology, any non-all-inclusive reality is "too small" to be God. Here is the definitive solution to the problem of a god "too small" for universal influence.

Sagan was correct in observing that the question of "convincing evidence" for the existence of a small god is "not significantly different" from the question of convincing evidence for the existence of intelligent extraterrestrials (108). The two questions are "not significantly different" insofar as both are about the existence of putative *parts* of reality. By contrast, where God is understood to be "the one all-inclusive *whole* of reality" (Ogden 1984: 21; italics added), the theological question is significantly different, and mere factual observations—such as seeing commandments on the moon or a crucifix in Earth orbit—cannot qualify as "absolutely clear-cut evidence" (165). More deeply empirical, logical, and metaphysical evidence is required. Though Sagan rightly identified the fact that the gods are "never seen" as an epistemological problem for affirming the existence of any small god (168), he did not consider the significantly different question of seeing the one reality large enough to be fully divine—the all-inclusive whole of reality.

For creatures that see, where anything real is seen, the all-inclusive whole of reality (God) is always "seen" in part and "never seen" in whole. Though seldom recognized, God is always seen in part where anything real is seen. No fragmentary creature (no part of reality) can ever see the comprehensive whole of reality. (Analogously, if single cells could see, no cell could see the whole body from inside the body of which it is a cellular part. At best, such a cellular part could see only some parts of that bodily whole.) At best, creaturely seeing always is seeing only some parts of the divine whole of reality.

Sagan's scientific search for God was seriously limited by his failure to search for the right kind of data concerning the all-inclusive whole of reality (the universal individual). Without attention to the living and inspiring whole of reality, Sagan's search for God was limited to searching for the "never seen" and "too small" god(s) of classical Western theology (putative part[s] of reality). In his search for God, Sagan was searching for life too small for universal influence.

ET as "Too Small"

Sagan's search for extraterrestrial life was also limited to searching for small life, that is, living individuals or communities of individuals that are small enough to live on a planet or moon. Nonetheless, the idea of searching for vastly larger individuals is suggested by Sagan's critical questions concerning the suspicious notion that the continuum of life stops at humans. Sagan said: "If there is . . . a continuum from self-reproducing molecules, such as DNA, to microbes, and an evolutionary sequence continuum from microbes to humans, why should we imagine that continuum to stop at humans? Why should there be an open-ended gap in the spectrum of beings? And isn't it a little suspicious that the gap would begin with us?" (103) Given a nonsuspicious reading of the small-to-large-to-larger continuum, the familiar pattern (many small individuals living within a large individual that lives within an even larger living individual) probably continues. Between earthly individuals and the one all-inclusive universal individual, there is plenty of space for multiple nestings of individuals that are both vastly more inclusive than earthly individuals and vastly less inclusive than the all-inclusive individual. (For example, a galaxy might be a living individual or part of a larger nonuniversal living individual or both. Certainly a galaxy is a very small part of the one universal/all-inclusive living individual.) Though Sagan did not advocate or even imagine searching for life forms that are vastly larger than us, his questions suggest the need for such searching. (Sir Fred Hoyle did conceive of astronomically larger-than-human intelligences, and he presented this conception in his science fiction book—*The Black Cloud* [1957].)

Science as Religious Experience

Sagan's "search for who we are" did not reveal that we are parts of an all-inclusive divine whole. Still, it did reveal much about who we are. We are at home on only one small planet. If we continue to commit "crimes against creation," we will become extinct much sooner than otherwise. We should repent of such crimes immediately. As explorers searching beyond our planet, who can look back at our "pale blue dot" (title of Sagan's 1994 book) from outer space, we should be inspired to have greater appreciation for our Earthly home and to have increased moral responsibility for its upkeep.

As indicated in the first chapter—"Nature and Wonder: A Reconnaissance of Heaven," exploring the cosmos inspires a sense of awe and "wonder" that Sagan recognized as a deeply religious sentiment. Sagan experienced scientific exploration of the cosmos as a form of worship. He held that "If a Creator God exists," surely such a God prefers "His votaries to admire the real universe," and, thus, "science is, at least in part, informed worship" (31). Even without recognizing God, Sagan and Druyan recognized the wonderful variety of scientific experience as a variety of religious experience.

9.3 Stephen W. Hawking and Leonard Mlodinow's *A Briefer History of Time* (New York: Banam Books, October 2005: 162 pages)

This critical review was published in the *Journal of the American Academy of Religion*, volume 74, number 4 (December 2006). With slight modifications, it is reproduced here with permission.

As the title indicates, *A Briefer History of Time* (BrHT) is a briefer (36 fewer pages) rendering of Stephen W. Hawking's original bestseller—*A Brief History of Time: From the Big Bang to Black Holes* (BHT) (Bantam Books, 1988). Originally stated "without mathematics in a form that people without a scientific education can understand," Hawking offered "basic ideas about the origin and fate of the universe" (BHT: vi). The same basic ideas—in many of the same words, lightly edited for easier reading, plus with new discussions of more recent developments in observation and theory, and all printed on glossy pages with 38 new colorful illustrations (minus some of the original content [autobiographical and technical])—yield *A Briefer History of Time* by Hawking and Mlodinow.

Unlike Hawking's 1988 brief history, the 2005 Hawking–Mlodinow briefer history does not include Carl Sagan's two-page Introduction. This is unfortunate because Sagan introduces an essential and important truth about the content of Hawking's book: "This is also a book about God . . . or perhaps about the absence of God" (BHT: x). Theological implications of cosmological models are much discussed in Hawking's brief history. Sagan notes, "The word God fills these pages" (BHT: x). There is no mention of God in the Foreword to the briefer history. Yet, because the briefer history maintains "the essential content of the original book" (BrHT: 2), the briefer history remains also about God.

The Hawking–Mlodinow briefer history also does not include Hawking's original three-page Acknowledgments section. This is unfortunate because here Hawking introduces a major feature of his work—that it has progressed from a "classical" phase to a "quantum" phase (BHT: vii). Then, in the first chapter Hawking explains that "the major theme of this book" is his continuing search for a new theory—a "quantum theory of gravity" (BHT: 12). In the briefer history, this major theme is not mentioned until part way into the third chapter.

During the "classical" phase, Hawking and Roger Penrose developed a cosmological model of the universe by employing a classical theory of gravity (Einstein's general theory of relativity). This model "implied that the universe must have a beginning, and possibly, an end" (BHT: 34). Our expanding universe is the result of one long ago event when space-time exploded into existence. This "big bang" is the ultimate historical "boundary." Before this initial moment of infinite space-time curvature and before this "big bang singularity," there was no universe, no space, no time, nothing. Hawking notes that classical big bang singularity theory is consistent with the idea that a divine Creator created the universe from nothing once upon a time—in the beginning (BHT: 46–47, 140–41).

During the "quantum" phase, Hawking changed his mind. Although he and Penrose employed a classical theory of gravity to develop the now widely accepted big bang singularity theory in 1970, since 1974, and especially since 1981, Hawking has favored a quantum theory of gravity (BHT: vii, 50, 115). A quantum theory of gravity is distinguished from a classical theory by including data from quantum mechanics. In classical theory, very large-scale interactions (planetary, stellar, and galactic) are understood to be governed by gravitational forces, and very small-scale interactions (atomic and subatomic) by quantum forces. Because our expanding universe was very small in much earlier times, Hawking figures quantum forces should be included. Thus, Hawking is now working to develop a "unified theory," a theory incorporating gravity (general relativity) and quantum mechanics, a quantum theory of gravity. Hawking and Mlodinow write, "We do not yet have such a theory, and we may still be a long way from having one, but we do already know many of the properties that it must have . . . we already know a fair amount about the predictions a quantum theory of gravity must make" (BrHT: 16).

A classical theory of gravity necessitates a universe beginning from a big bang singularity. A quantum theory of gravity allows for a universe with "no singularity at the beginning," a universe with "no beginning," an eternal universe with "no boundary" (BHT: 50, 116). Hawking holds that his theory of a universe with no boundary has "profound implications for the role of God" as Creator (BHT: 140, 174; BrHT: 141). For Hawking, "no beginning, no moment of Creation" implies that there is no clear "role" for a divine Creator

(BHT: 116, 136, 140–41, 172–74; BrHT: 102–3, 134–35, 140–42), "nothing for a Creator to do" (BHT: x).

In addition to explicating the theological implications of classical and quantum cosmological models, Hawking identifies implications for scientific inquiry. Classical singularity theory "predicts its own downfall" by requiring that all scientific theories "break down at the big bang" (BrHT: 68, 102; BHT: 148). No science can reach this boundary. By contrast, given a quantum theory of gravity and a universe with no boundary, the "laws of science" would not be required to fail at some boundary (BrHT: 85, 103). A successful quantum theory of gravity would prevent the necessary failure of scientific inquiry and avoid need for "appeal to God or some new law to set the boundary conditions for space-time" (BrHT: 103).

Hawking's reasoning about theological implications is correct as far as it goes, but it does not go far enough to be adequate. Hawking reasons correctly in holding that accepting his nonclassical cosmology implies rejecting a notion of God as the once-and-only-once-upon-a-time (in the beginning) Creator. Hawking, however, shows no awareness that there are alternative conceptions of God as Creator, including a conception Hartshorne calls "neoclassical." According to neoclassical thinking, proper religious language about divine creativity is *not* about only one event (the beginning); instead, it is about divine influences upon all events at all times in all spaces. Moreover, Hartshornean neoclassical thought denies that the idea of an absolute beginning of time makes sense. Insofar as Hawking considers only selected aspects of a classical view of God, his account of the theological implications of cosmological models is unavoidably fatally inadequate.

Collaborating with Mlodinow did not yield a more adequate account of theological implications or a significant advance in cosmological theory. This collaboration did succeed, however, in making scientific cosmology more easily available. In addition to being briefer, less technical, and "more leisurely" (BrHT: 1), the briefer history includes references to new data from the Hubble Space Telescope and the COBE satellite and discussion of new increments in progress toward a unified theory. Casual readers are likely to find the new chapter on "Wormholes and Time Travel" more entertaining than the original chapter called "The Arrow of Time." And for such readers, it is probably good that the briefer history omits the original discussion of "imaginary numbers," "imaginary time," and the "three arrows" of time. Readers looking for a theory about "Black Holes," and why Hawking argues "Black Holes

Ain't So Black," will need to read the earlier history (BHT: Chapters 6 and 7). The new illustrations are attractive and helpful.

For leisure reading, the briefer history can replace the brief history. For scholarly work, the briefer history is an addendum, a review, and update. Either way, we should be very grateful to Hawking and Mlodinow for sharing the content and excitement of recent cosmological inquiry by translating math-based theory into relatively simple words and pictures.

9.4 Larry Witham's *The Measure of God: Our Century-Long Struggle to Reconcile Science and Religion.* New York, New York: HarperCollins, 2005: 358 + ix pages (hardcover).

A previously unpublished critical review.

Larry Witham's *The Measure of God* (2005) is about scientific natural theology and the Gifford Lectures. By telling the story of Gifford lecturers struggling to measure and construct scientific language about God, award-winning religious news writer and book author Witham illuminates "our century-long struggle to reconcile science and religion." Plus, he offers his own judgment that theism, though "not proven," is "far more plausible" than atheism (282).

Witham's history is also about the influences of "metaphysical Scotland" (in contrast to "utilitarian England") upon North American thought. Under the continuing influence of the "Scottish Enlightenment" (1740–90), the Gifford Lectures began when Scottish Lord Adam Gifford (b.1820, d. 1887) in his 1885 will endowed a perpetual lecture series at four Scottish universities: Aberdeen, Edinburgh, Glasgow, and St. Andrews.

Lord Gifford was seeking to encourage "lively and perpetual debate on science and 'all questions about man's conception of God or the Infinite'" (1). He was hugely successful. The first Gifford Lecture was delivered at Glasgow in 1888. Except for 1941–46 when there were no Gifford Lectures, the Giffords embrace the whole twentieth century and more. As of 2005, the year of publication, 220 Gifford lecturers had published more than 200 Gifford Lectures as books. Many of these books are famous classics.

Gifford lecturers include many of our most influential philosophers, scientists, theologians, ethicists, and other scholars. They include Frederich Max Müller, James Hutchison Stirling, Josiah Royce, William James, James George Frazer, Henri Bergson, Alfred North Whitehead, John Dewey, Albert Schweitzer, Karl Barth, Reinhold Niebuhr, Emil Brunner, Niels Henrik David Bohr, Charles Earle Raven, Arnold Joseph Toynbee, Paul Tillich, Charles Arthur Campbell, Rudolf Bultmann, Werner Carl Heisenberg, Hannah Arendt, Alfred J. Ayer, Owen Chadwick, Stanley L. Jaki, Basil G. Mitchell, John C. Eccles, Ninian

Roderick Smart, Iris Murdoch, Freeman J. Dyson, John Macquarrie, Jürgen Moltmann, Paul Ricoeur, Carl Sagan, John Hick, Anthony Flew, Alvin Plantinga, Alasdair MacIntyre, Raimundo Panikkar, Ian Barbour, Jaroslav Pelikan, Martha Craven Nussbaum, Arthur Robert Peacocke, Roger Penrose, John Polkinghorne, Nicholas Wolterstorff, Charles Taylor, David Tracy, Ralph McInerny, Mohammed Arkoun, Stanley Hauerwas, Peter van Inwagen, Stephen Toulmin, Noam Chomsky, and others. Accordingly, Witham judges that the Gifford Lectures are "perhaps the best repository on earth for this discussion" of science and theology (vii).

Given 220 Gifford Lecturers, and more coming, the whole story cannot be told. Though all of the lecturers (from 1888 to 2005) are listed in the Appendix, Witham discusses "only a fraction . . . the most famous . . . and those who illustrate broader trends" (vii). And because he is writing for North American readers, Witham highlights "the American greats, although, by far, more characters in this story are European and, for that matter, nearly all men" (vii). Witham simplifies the story and illustrates broad trends by describing selected lectures as contributions to a historical drama with four partly overlapping acts.

This four-act drama is artfully connected to contextual and biographical data, such as Adam Gifford serving as chief prosecutor in an infamous murder case—"The Standyford Mystery" (Chapter 1). Witham frames the whole drama by references to great works of graphic art, including "Ancient of Days" (1794) by Englishman William Blake (reproduced on the paper jacket) and "Revered Walker Skating on Duddingston Loch" (1795) by Scotsman Sir Henry Reburn. "As a student and lawyer, Gifford had been an avid ice skater down at Duddingston Loch" (15).

Act one, starting with Gifford and the early Gifford Lectures (from 1888 to 1900), is about the decline of philosophical idealism and the rise of materialism, especially materialist modern science. According to Chapter 2, "The End of Philosophy: The Rise and Fall of Idealism," this period marked "the end of philosophy, which speculated about God, and the dawning of natural science, which asked about God amid the material facts of the world" (27). Witham shows that the transition from the dominance of idealism to the dominance of materialism is exhibited in the Giffords. Though "idealist philosophers swamped the early choices for Gifford lecturers" (40), in subsequent years, Darwinian materialists "would sweep the field" and "declare the end of philosophy" (45).

Act two, taking place mostly between 1901 and World War I, marks the beginnings of modern scientific inquiries into questions about God and religion. Accordingly, in act two, "the sciences became major contributors to the Gifford Lectures" (57). There is a chapter for each of five scientific fields: anthropology (Chapter 3), psychology (Chapter 4), physics (Chapter 5), sociology (Chapter 6), and history (Chapter 7). Though the general trend among scientists was to reduce reality to insentient energy and matter (reductionist materialism), act two included a "small revolution" (98)—the emergence of an organic (nonreductionist) account of reality that eventually came to be called "*process philosophy*" (100; italics added).

This "small revolution" was inaugurated in the Giffords by William James's *The Varieties of Religious Experience: A Study of Human Nature* (1901 and 1902 at Edinburgh). (Witham observes that in *The Principles of Psychology* [1891], William James introduced "radical empiricism" [83].) This small revolution was advanced in other Gifford Lectures, including Hans Driesch's *The Science and Philosophy of the Organism* (1907 and 1908 at Aberdeen); James Ward's *The Realm of Ends, or Pluralism and Theism* (1907–9 at St. Andrews); Henri Bergson's *The Problem of Personality* (1914 at Edinburgh) (also note Bergson's 1907 *Creative Evolution*); John Arthur Thomson's *The System of Animate Nature* (1914–16 at St. Andrews); Samuel Alexander's *Space, Time and Deity* (1917 and 1918 at Glasgow); Conwy Lloyd Morgan's *Emergent Evolution and Life, Mind, and Spirit* (1921–23 at St. Andrews); and Alfred North Whitehead's *Process and Reality: An Essay in Cosmology* (1927 and 1928 at Edinburgh).

Act three, taking place mostly between World War I and World War II, is about two movements contrary to natural theology. According to Chapter 8—"From Barth to Being: The Revolt against Reason"—the first movement (from Barth) was a Protestant theology granting exclusive favor to Christian revelation. The second movement (to Being) emphasizing subjective individual experiences and freedoms was "existentialism."

Karl Barth led the first of these two movements. Barth's Gifford Lectures, *The Knowledge of God and the Service of God According to the Reformation: Recalling the Scottish Confession of 1560* (1937 and 1938 at Aberdeen), included a declaration of opposition to natural theology. A challenge to Barth's account of natural theology came from James Barr's *Biblical Faith and Natural Theology* (1990 and 1991 at Edinburgh). A vigorous defense of Barth's evaluation of natural theology came from Stanley Hauerwas's *With the Grain of the Universe: The Church's Witness and Natural Theology* (2000 and 2001 at St. Andrews). Hauerwas, writes Witham, turned the Barth debate "on its head" by declaring Barth "the true natural theologian" because Barth saw that natural theology requires "a specific God . . . a trinitarian Creator who worked in the world by way of Christ and the Holy Spirit in the church" (215).

Leading figures in the second movement (existentialism) were Søren Kierkegaard, Wilhelm Dilthey, Edmund Husserl, Martin Heidegger, and later Jean Paul Sartre. Rudolf Bultmann is "the father of a biblical form of existentialism" (220). Bultmann's *History and Eschatology: The Presence of Eternity* (1955 at Edinburgh) included arguments for "demythologizing" and "realized eschatology" in service to an "existential encounter with God in Christ" (219–20). Witham identifies other Gifford lecturers exhibiting existentialist influences: Etienne Gilson (1931 and 1932 at Aberdeen), Jacques Maritain (whose Giffords were cancelled during World War II), Gabriel Marcel (1949 and 1950 at Aberdeen), Paul Tillich (1953 and 1954 at Aberdeen), and John Macquarrie (1983 at St. Andrews). Witham argues, "No one took Being to its theological conclusion more completely than the Protestant existentialist Paul Tillich" for whom God is the "ground of being" and Christ is the "new being" (220, 221).

Act four, spanning the post-World War II period, is described as "a happy time for natural theology" (6). Act four is covered in Chapters 9—"A Designer Universe: God and the New Science"—and 10—"Religious Pluralism: The Limits of Knowledge." This act focuses upon the support natural theology received from a renewed appreciation of the "small revolution" in act two and from new scientific discoveries, including evidence of "a designer universe" in the form of finely tuned cosmic constants seemingly designed to permit human emergence. Responding to pluralism, especially religious pluralism, is another feature of act four.

Act four was "inaugurated" by Charles Earle Raven's *Natural Religion and Christian Theology* (1950–52 at Edinburgh). Raven identified new scientific discoveries, a new awareness of the limits of science, and a "desire to make science moral" as factors that "reinvigorated" the interchange between science and theology (230). Moreover, Raven predicted the breakdown and rejection of reductionist-materialist accounts of nature, and he predicted Whiteheadian process thought would become more influential.

Witham's analysis of subsequent Gifford Lectures shows Raven predicted correctly. Arthur Robert Peacocke's *Nature, God and Humanity: Towards a Christian Theology for a Scientific Age* (1992 and 1993 at

St. Andrews) and Holmes Rolston III's *Genes, Genesis and God: Values and Their Origins in Natural and Human History* (1997 and 1998 at Edinburgh) are antireductionist. Alister Clavering Hardy's *Living Stream* and *The Divine Flame* (1963 and 1965 at Aberdeen), Freeman Dyson's *Infinite in All Directions* (1983–85 at Aberdeen), and Ian Barbour's *Religion in an Age of Science* and *Ethics in an Age of Technology* (1989 and 1990 at Aberdeen) exhibit process influences. Hardy, a biologist who saw process metaphysics as consistent with neo-Darwinism, recognized Charles Hartshorne as "the greatest theological proponent of process metaphysics" (247). Witham adds, "In the opinion of many, he [Hartshorne] should have been appointed a Gifford lecturer long before his death in 2000—at the age of 103" (247).

Process thought, though increasingly influential, is not embraced by all act-four Gifford lecturers. Witham sees "a deep aversion" to process thought in Eric Lionel Mascall's *The Openness of Being: Natural Theology Today* (1970 and 1971 at Edinburgh) and in Stanley Jaki's *The Road of Science and the Ways of God* (1974–76 at Edinburgh) (256–57). Dualism lives in Sir John C. Eccles' *The Human Mystery* and *The Human Psyche* (1979 at Edinburgh). And John Polkinghorne's *The Faith of a Physicist: Reflections of a Bottom-Up Thinker* (1993 and 1994 at Edinburgh) favors classical Christian doctrines. Pluralism prevails throughout act four.

In Chapter 11—"The Measure of God: Scotland, America, and the Giffords"—Witham appreciates Scottish influences. In addition to appreciating the enduring influences of Scottish Lord Gifford—"patron saint" for natural theology (299)—Witham finds a North American account of Scottish influences in Nicholas Wolterstorff's *Thomas Reid and the Story of Epistemology* (1995 at St. Andrews). Scottish philosophers Thomas Reid (b. 1710, d. 1796) and Francis Hutcheson (b. 1694, d. 1746; founder of the Scottish Enlightenment) developed notions of "common sense" and "self-evident truth" that influenced colonial Americans, including Thomas Paine and Thomas Jefferson (285–86). Moreover, Reid's work influenced American pragmatists Charles Sanders Peirce and William James. Accordingly, Witham quotes Wolterstorff as saying, "American pragmatism is the successor of Scots common sense philosophy" (289).

Witham also acknowledges Scottish influences upon his own thinking. His judgment that theism is "not proven" is in agreement with Scottish Gifford lecturer John Laird's *Theism and Cosmology* and *Mind and Deity* (1938–40 at Glasgow). Witham notes that Scottish law adds "not proven" to the list of possible verdicts and that Lord Gifford was a Scottish judge. Furthermore, Reid's "Scottish Philosophy" inspires

Witham to conclude that twenty-first-century natural theologians should revisit "common sense" and "self-evident" truth (7, 283).

The Appendix—"The Gifford Lectures, 1888 to 2005"—is a 13-page chronological list of the Gifford Lecturers. It shows full names, specific lecture years, universities where the lectures were delivered, lecture titles, birth-death years, and the academic positions of lecturers at the time of their lectures. Regrettably, this data is not referenced in the Index.

Witham's four-act dramatization simplifies helpfully an extremely complex history and illuminates broad trends. This is a very important contribution to the history of natural theology.

9.5 John W. Cooper's *Panentheism—The Other God of the Philosophers: From Plato to the Present*. Grand Rapids, Michigan: Baker Academic: November 2006. 358 pages. $34.99 (hardcover).

A previously unpublished critical review.

The first word of the title—"Panentheism"—can be misleading for those of us looking for a book-length deliberation on divine all-inclusiveness. Cooper's book is about much more. A title more fully descriptive of the book's content would be "*A historical survey* ['From Plato to the Present'] *of a vast range of reason-based relational alternatives to supernatural-revelation-based classical theism, followed by brief arguments for the conviction that classical theism is a more adequate expression of Christian faith than any of the relational alternatives.*"

The remainder of Cooper's title—"The Other God of the Philosophers"—is precisely on point. Cooper seizes the "rhetorical advantage" from those critics of classical theism that refer to classical theism's conception of God as "the God of the philosophers" on account of classical theism's indebtedness to classical Greek philosophers. He argues that those presenting relational alternatives to classical theism are also indebted to classical Greek philosophers, and therefore any one of the relational conceptions of God can properly be called "the *other* God of the philosophers."

Relational alternatives to classical theism represent God as "a God who is involved in time, interacts with creatures, and is affected by them" (15). In contrast to such "relational views of God" now "endorsed by a large majority of theologians along a broad spectrum from religious pluralism, on one end, to evangelical Christianity, on the other" (15–16), classical theism favors a nonrelational view of God. Cooper explains:

> Western classical theism asserts that God in himself is maximal Being—absolutely self-sufficient, eternal, immutable, omnipotent, omniscient, completely active, and most excellent in every way. Although he does not need the world, God eternally and freely chooses to create it from nothing and sustain it through time. He is immanent in the sense that he

> is supernaturally present to all beings and events at all times and places throughout the history of the world, empowering creatures and effectuating his eternal knowledge and will through their natural existence and free actions. But God in himself is utterly transcendent, all-determining, and changeless. The world is not part of his nature or existence. He does not exist in time or as part of the cause-and-effect networks in terms of which creatures exist and relate within the world order. Nothing temporal affects his existence, knowledge, or will. In other words, classical theism affirms that God is eternal and immutable even in relationship with his creation. (14–15)

Cooper's historical survey includes descriptions of relational alternatives to classical theism—alternatives ranging from "minor modifications of classical theism to major revisions of classical theism, to varieties of panentheism, to new versions of naturalistic pantheism" (17)—presented by more than 70 individuals. The 70 listed in the table of contents are Plato, Pseudo-Dionysius, John Scotus Eriugena, Meister Eckhart, Nicholas of Cusa, Jakob Böhme, Giordano Bruno, Baruch Spinoza, Jonathan Edwards, Gotthold Lessing, Gottfried Herder, Friedrich Schleiermacher, Immanuel Kant, Johann Gottlieb Fichte, Friedrich Wilhelm Joseph von Schelling, Georg Friedrich Wilhelm Hegel, Karl Krause, Isaak Dorner, Gustav Fechner, Hermann Lotze, Otto Pfleiderer, Ernst Troeltsch, Samuel Taylor Coleridge, Thomas Hill Green, John and Edward Caird, James Ward, Andrew Seth Pringle-Pattison, Samuel Alexander, William Ralph Inge, Ralph Waldo Emerson, Charles Sanders Peirce, William James, Jules Lequier, Charles Renouvier, Henri Bergson, Teilhard de Chardin, Alfred North Whitehead, Charles Hartshorne, John B. Cobb Jr., David Ray Griffin, Paul Tillich, Martin Heidegger, Han-Georg Gadamer, Nicolai Berdyaev, William Temple, John A. T. Robinson, John Macquarrie, Karl Rahner, Hans Küng, Martin Buber, Muhammed Iqbal, Sarvepalli Radhakrishnan, Masao Abe, Alan Watts, Miriam Starhawk, Jürgen Moltmann, Wolfhart Pannenberg, James H. Cone, Gustavo Gutiérrez, Juan Luis Segundo, Leonardo Boff, Rosemary Ruether, Sallie McFague, Matthew Fox, Ian Barbour, Paul Davies, Arthur Peacocke, Philip Clayton, and John Polkinghorne (7–8). Some of these individuals are explicit panentheists. Most are not. The common denominator among them is advocacy of a "relational alternative to classical theism," not explicit panentheism—not divine all-inclusiveness. Hence, another more fully descriptive title for this book might be "*Relational Theism—The Other God of the Philosophers: From Plato to Polkinghorne.*"

In the introductory chapter, Cooper notes that the literal meaning of the term "panentheism" is "all-in-God-ism," that "panentheism" derived from a translation of "*Allingottlehre*," a term coined by Karl Krause (b. 1781, d. 1832), and that since its mid-twentieth-century advance by Charles Hartshorne, panentheism has acquired "a commonly accepted generic definition" that Cooper renders as "God and the world are ontologically distinct and God transcends the world, but the world is in God ontologically" (26–27). Nonetheless, because "there are widely differing ways of understanding panentheism," introducing panentheism is "complicated" (27). "For the sake of clarity," Cooper identifies five sets of distinctions: "explicit and implicit panentheism; personal and nonpersonal panentheism; part-whole and relational panentheism; voluntary and/or natural panentheism; and classical (divine determinist) or modern (cooperative) panentheism" (27).

Also in the introductory chapter, Cooper specifies two purposes for his book. The first and main purpose is to offer a survey of "the development and proliferation of panentheism from its roots in Neoplatonism to its many different branches in the twenty-first century" (19). The second purpose is to offer "a critical and apologetic response to panentheism" (20). Chapters 2 through 13 serve the first purpose. The fourteenth and final chapter serves the second purpose. The intended readers are students needing a historical introduction to "the great panentheist tradition" (22).

The survey chapters are: "Panentheism from Plato through Christian Neoplatonism" (Chapter 2), "Pantheism and Panentheism from the Renaissance to Romanticism" (Chapter 3), "Schelling and Hegel: The Godfathers of Modern Panentheism" (Chapter 4), "Nineteenth-Century Proliferation" (Chapter 5), "Teilhard de Chardin's Christocentric Panentheism" (Chapter 6), "Process Theology: Whitehead, Hartshorne, Cobb, and Griffin" (Chapter 7), "Tillich's Existential Panentheism" (Chapter 8), "Diversity in Twentieth-Century Philosophy, Theology, and Religion" (Chapter 9), "Moltmann's Perichoretic Panentheism" (Chapter 10), "Pannenberg's Panentheistic Force Field" (Chapter 11), "Panentheistic Liberation and Ecological Theologies" (Chapter 12), and "Panentheism in Theological Cosmology" (Chapter 13).

In the final chapter—"Why I Am Not a Panentheist"—Cooper argues that "classical theism is more adequate for providing a biblically faithful, philosophically sound articulation of Christian theology, salvation history, and the Christian worldview" (342). Cooper endorses "Reformed Christian classical theism."

Three Philosophical Issues: Divine Freedom, Mind–Body Analogy, and Proof from Infinity

In a section of the final chapter called "Philosophical Issues," Cooper identifies "three representative philosophical topics on which panentheists claim the high ground: God's freedom, the mind-body relation as a model for the God-world relation, and the panentheistic proof from the true Infinite" (337). There are subsections addressing these "high ground" topics.

In subsections called "Divine Freedom" and "God's Self-Sufficiency and Freedom to Create," Cooper takes issue with panentheist accounts of God's freedom to create. Panentheists hold a compatibilist view of God's freedom to create. On this view, God must create creations and creative creatures. Cooper judges that this "limits" divine freedom by not allowing God freedom to create nothing. Thus, divine freedom is "an oxymoron in almost all panentheism" (326). Classical theists hold to a libertarian view of God's freedom to create. On this view, God is free to choose to create "nothing at all" (325). Because generic panentheism does not require a compatibilist view, Cooper concludes that "to the almost universal extent that panentheists hold it, they seem philosophically arbitrary, confused, or incoherent" (337).

Unfortunately, the coherence of language about creating nothing is more assumed than critically considered. Talking about a living Creator (or creature) "creating nothing at all" has no obvious meaning. And obviously, "noncreative Creator" is self-refuting. According to Hartshorne, rejecting such language does not limit divine freedom. Instead, it contributes to limiting our language to meaningful constructions.

In a subsection called "The Mind-Body Analogy for God and the World," Cooper argues that "the mind-body model for God and the world is philosophically inadequate because it is a poor analogy, at least for high personal panentheism" (338). Cooper does not identify a richer analogy, and he offers no critical deliberation on analogy as such. He says, "classical theism's view—that God is a purely spiritual, bodiless being who is universally present to creatures other than himself—seems to do a better job" (339).

In a subsection called "The Proof from Infinity," Cooper argues that "Nicholas of Cusa, Hegel, Hartshorne, Pannenberg, Clayton, and others allege a philosophical proof for their theology—the argument from Infinity" (339). Cooper concedes that "the argument from infinity is sound in a formal sense," because "in some sense, nothing can be 'outside' God" (339). Cooper argues, however, that this "does not prove panentheism," and he concludes that "panentheism is not entailed by the argument from Infinity" (339).

Cooper is correct. As we shall soon argue, infinity does not entail panentheism. He is incorrect, however, in counting Hartshorne among those who argue from infinity to panentheism, incorrect in conceding that the argument from infinity is formally sound, and incorrect in thinking that his correct view (that infinity does not entail panentheism) is contrary to panentheism as such.

The high ground explicitly claimed by panentheism is divine all-inclusiveness. Divine all-inclusiveness defines panentheism and distinguishes it from classical theism. Classical theists hold that God is a "purely spiritual" (339) part of reality. Panentheists hold that God is the all-inclusive whole of reality. Thus, where classical theism is contingently monotheistic (see 4.13 in this book), panentheism is necessarily monotheistic because logic admits only one all-inclusive reality. Moreover, divine all-inclusiveness—or, to say the same, divine "surrelativism"—distinguishes panentheism from other relational alternatives (less than all-inclusive and less than "supremely-relative" conceptions of the divine) such as polytheism and pantheism. Unfortunately, "all-inclusive" does not appear in Cooper's subject index. By contrast, "Infinite" has ten references, "Infinite Being" has two, and "infinity" has 19 (355).

Infinity is not clearly distinguished from all-inclusiveness in "the proof from infinity." Cooper's account of that proof is this: "Because God is absolutely infinite, nothing can be completely other or outside him. For, if anything were, then God would be limited by it, that is, finite, that is, not-infinite, which is impossible by definition" (339). In contrast to this "proof from infinity," Hartshornean panentheists hold that because God is the all-inclusive whole of reality, by definition, nothing real can be outside God. All-inclusive wholeness, not infinity, is the definition by which "outside God" is impossible.

Cooper goes on to assert, "Panentheists construe infinity in terms of ontological 'in-ness'" (339). No doubt, some panentheists do construe infinity as all-inclusiveness. Hartshornean panentheists, however, recognize that taking the infinite to be the all-inclusive is a mistake in theology and in mathematics. One million, two million, three million, four million, and so on is infinite, but not all-inclusive. Distinguishing infinity from all-inclusiveness makes it more obvious that infinity does not entail panentheism, not even "in a formal sense."

Conclusion

Cooper is successful in neutralizing any "rhetorical advantage" gained by equating classical theism with the "God of the philosophers." On this point, he is in full agreement with Whitehead and Hartshorne. As Cooper observes, Whitehead describes Western philosophy as footnotes to Plato, and Hartshorne acknowledges his debt to Plato and classical Greek philosophy by calling his metaphysical thought "neoclassical" (181).

Cooper is correct in judging that, as an articulation of biblical revelation and "fully trinitarian" Christian doctrines, most versions of panentheism are "less than adequate" (324). Where "inadequate" means "this is correct as far as it goes, but it does not go far enough to fully serve some robust purpose," Cooper's judgment is consistent with metaphysical panentheism. To be sure, Hartshorne insists that metaphysical statements include no contingent-factual data. Hence, necessarily, metaphysical panentheism is never adequate to any religion (which is by definition a partly historical-empirical phenomenon).

With respect to his three philosophical issues, Cooper fails to take the high ground claimed by Hartshornean panentheism. The panentheist attempt to restrict language about divine creativity to meaningful constructions is mistaken for a limit on divine freedom. The mind–body analogy is not improved upon at all. And "proof from infinity" is mistaken for a formally sound and defining feature of panentheism.

As introduced in Cooper's historical survey, the "great panentheist tradition" covers a broad range of relational alternatives to classical theism. If Cooper had focused more narrowly on divine all-inclusiveness (the defining feature of panentheism), Hartshorne would have been central, instead of one among 70 or so mostly nonexplicit panentheists. Though not tightly focused on divine all-inclusiveness, Cooper's book is an illuminating historical survey of many various relational alternatives to classical theism. And, as such, it is a valuable resource for the history of theological doctrines.

9.6 Autobiographical Astro-Theology

(This autobiographical statement, originally titled "The Heavens Declare the Glory," was published in *Perspective: A Publication by the Perkins School of Theology* [2008] and is reproduced here with permission.)

I was born during winter solstice in Greensboro, North Carolina, U.S.A. in the year 1953. My earliest memories are of families and churches in rural Woodsdale near Roxboro, North Carolina.

I remember being a toddler traveling late at night along a lonely country road in an automobile overloaded with churchgoers returning from worship. The car stalled, and came to a stop. We had to walk. I remember riding on my father's shoulders as he walked with my mother and other walkers. And I remember leaning backwards, hanging upside down, as my father held me securely by my ankles, and looking upwards into a dark starry night sky, and listening to the short line of walkers who were singing hymns and spirituals while walking. For me, that was a joyful adventure.

From childhood through high school graduation in 1972, my favorite toys included microscopes, telescopes, star charts, shortwave radios, amateur radio transmitters (ham radios), and telegraph keys. While a student at Salisbury High School in Salisbury, North Carolina, I presented a science-fair paper at the University of North Carolina at Chapel Hill, and again at the University of London, concerning radio astronomy: "The Radio Sun as Seen with a Home-Made Radio Telescope (a two-meter interferometer with phased Yagi-Uda arrays)" (1971–72, unpublished).

Though my present academic work is about theology and theological ethics, I continue to be interested in night skies. And just in case my car stalls during night travels between Texas and North Carolina, I habitually travel with binoculars (and sometimes a telescope in the trunk) because, as always, "the heavens are declaring the glory of God" (Psalm 19:1 [NRSV]).

Bibliography

Abraham, William J. (2015). "Brain Battles: Theology and Neuroscience" in *A Man of Many Parts: Essays in Honor of John Westerdale Bowker on the Occasion of His Eightieth Birthday*, edited by Eugene Lemcio. Eugene: Pickwick.

Adler, Mortimer J. (1992). "Natural Theology, Chance, and God" in *Great Ideas Today*. Chicago: Encyclopedia Britannica; plus "Response to Mortimer Adler" by Owen Gingerich; also www.thegreatideas.org.

Albrecht, A. and P. J. Steinhardt (April 26, 1982). "Cosmology for Grand Unified Theories with Radiatively Induced Symmetry Breaking" in *Physical Review Letters*, vol. 48, 1220–23 [doi:10.1103/PhysRevLett.48.1220].

Alexander, Conel M. O'D. (2011). "A Common Origin for Organics in Meteorites and Comets: Was It Interstellar?" in *Molecular Universe: Proceedings of the International Astronomical Union Symposium Number 280*, edited by José Cernicharo and Rafael Bachiller. Paris: International Astronomical Union [doi:10.1017/S1743921311025051].

Alexander, Conel M. O'D. with R. Bowden and others. (March 19, 2012). "The Origin of Water in Chondrites and Volatiles in the Terrestrial Planet Region" [#1929] presented at *43rd Lunar and Planetary Science Conference*, March 19–23, Woodlands, Texas.

Alexander, Conel M. O'D. with R. Bowden and others. (August 10 [July 12], 2012). "The Provenances of Asteroids, and Their Contributions to the Volatile Inventories of the Terrestrial Planets" in *Science*, vol. 337, issue 6095, 721–23 [doi:10.1126/science.1223474].

Alexander, Conel M. O'D. with G. D. Cody and others. (2008) "Organics in Meteorites—Solar or Interstellar?" in *Organic Matter in Space: Proceedings of the 251st Symposium of the International Astronomical Union*, February 18–22, Hong Kong, China, edited by Sun Kwok and Scott A. Sandford. Cambridge: Cambridge University Press.

Alexander, Conel M. O'D. with M. Foge and others. (July 17, 2007). "The Origin and Evolution of Chondrites Recorded in the Elemental and Isotropic Compositions of Their Macromolecular Organic Matter" in *Geochimica et Cosmochimica Acta*, vol. 71, 4380–403.

Alexander, Mabel Hovdahl. (2004). *Via Oklahoma: And Still the Music Flows*. Oklahoma City: Oklahoma Heritage Association.

Alfvén, Hannes. (1966). *Worlds-Antiworlds: Antimatter in Cosmology*. San Francisco: Freeman.
Alfvén, Hannes. (1969). *Atom, Man, and the Universe: The Long Chain of Complications*, translation by John Hoberman. San Francisco: Freeman.
Alfvén, Hannes. (1973). *On the Origin of the Solar System*. Westport: Greenwood.
Alfvén, Hannes. (1977). "Cosmology: Myth or Science?" in *Cosmology, History, and Theology* [from Third International Colloquium at University of Denver, November 5–8, 1974], edited by W. Yourgrau and A. D. Breck. New York: Plenum.
Alfvén, Hannes. (1981). *Cosmic Plasma*. Dordrecht: Kluwer.
Alfvén, Hannes. (February 1990). "Cosmology in the Plasma Universe: An Introductory Exposition" in *IEEE Transactions on Plasma Science*, vol. 18, 5–10.
Alfvén, Hannes and Kerstin Alfvén. (1972). *Living on the Third Planet*, translation by Eric Johnson. San Francisco: Freeman.
Alfvén, Hannes and Gustaf Arrhenius. (1976). *Evolution of the Solar System*. Washington: NASA.
Alfvén, Hannes and Asoka Mendis. (April 21, 1977). "Interpretation of Observed Cosmic Microwave Background Radiation" in *Nature*, vol. 266, issue 5604, 698–99 [doi:10.1038/266698a0].
Allen, David A. and Dayal Tissa Wickramasinghe. (November 19, 1981). "Diffuse Interstellar Absorption Bands between 2.9 and 4.0 μm" in *Nature*, vol. 294, issue 5338, 239–40 [doi:10.1038/294239a0].
Allen, D. A. and D. T. Wickramasinghe. (October 21, 1987). "Discovery of Organic Grains in Comet Wilson" in *Nature*, vol. 329, issue 6140, 615–16 [doi:10.1038/329615a0].
Allen, Paul. (2006). *Ernan McMullin and Critical Realism in the Science-Theology Dialogue*. Aldershot: Ashgate.
Alpher, Ralph A. (Winter 1983). "Theology of the Big Bang" in *Religious Humanism*, vol. 17, issue 1, 1–13.
Alpher, Ralph A. with H. Bethe and George Gamow. (April 1 [February 18], 1948). "Origin of Chemical Elements" [αβγ paper] in *Physical Review*, vol. 73, issue 7, 803–4 [doi:10.1103/PhysRev.73.803].
Alpher, Ralph A. and Robert C. Herman. (November 13, 1948). "Evolution of the Universe" in *Nature*, vol. 162, 774–75 [doi:10.1038/162774a0].
Alpher, Ralph A. and Robert C. Herman. (December 15 [September 1], 1948). "On the Relative Abundance of the Elements" in *Physical Review*, vol. 74, issue 12, 1737–42 [doi:10.1103/PhysRev.74.1737].
Alpher, Ralph A. and Robert C. Herman. (April 1, 1949 [December 27, 1948]). "Remarks on the Evolution of the Expanding Universe" in *Physical Review*, vol. 75, issue 7, 1089–99 [doi:10.1103/PhysRev.75.1089].
Alpher, Ralph A. and Robert C. Herman. (April 1, 1950). "Theory of the Origin and Relative Abundance Distribution of the Elements" in *Review of Modern Physics*, vol. 22, issue 2, 153–212 [doi:10.1103/RevModPhys.22.153].
Alpher, Ralph A. with Robert C. Herman and G. Gamow. (November 1, 1948 [September 15, 1948]). "Thermonuclear Reactions in the Expanding

Universe" in *Physical Review*, vol. 74, issue 9, 1198–99 [doi:10.1103/PhysRev.74.1198.2].

Alpher, Ralph A. with Robert C. Herman and G. Gamow. (December 1967). "Thermal Cosmic Radiation and the Formation of Protogalaxies" in *Proceedings of the National Academy of Sciences*, vol. 58, issue 6, 2179–86.

Altwegg, K. with H. Balsiger and others. (December 10 [October 1], 2014). "67P/Churyumov-Gerasimenko, a Jupiter Family Comet with a High D/H Ratio" in *Science Magazine*, 1–6 [doi:10.1126/science.1261952].

Ames, Roger and David Hall. (2001). *Focusing the Familiar: A Translation and Philosophical Interpretation of the Zhongyong*. Honolulu: University of Hawaii Press.

Anselm [Saint Anselm of Canterbury (1033–1109)]. (1995 [1977]). "Proslogion" (1077–78), translation by T. Williams, in *Classics of Western Philosophy*, fourth edition, edited by Steven Cahn. Indianapolis: Hackett.

Arca, Frank. (2007). *The Physical Origin of the Universe*. Lincoln: iUniverse.

Ariatti, Annalisa and Paul Comtois. (1993). "Louis Pasteur: The First Experimental Aerobiologist" in *Aerobiologia*, vol. 7, issue 9, 5–14.

Arnett, David. (2011 [2005]). "Sir Fred Hoyle and the Theory of the Synthesis of the Elements" in *The Scientific Legacy of Fred Hoyle*, edited by D. Gough, foreword by M. Rees. Cambridge: Cambridge University Press.

Arnould, Jacques. (May 2010). "Purposeful Panspermia: The Other Conquest of Space? Ethical Considerations" in *Journal of Cosmology*, vol. 7, issue 6, 1726–30.

Arp, Halton C. (1987). *Quasars, Redshifts and Controversies*. Berkeley: Interstellar Media.

Arp, Halton C. (2003). *Catalogue of Discordant Redshift Associations*. Montreal: Apeiron Academic.

Arp, Halton C. with Geoffrey Burbidge, Fred Hoyle, J. V. Narlikar, and N. C. Wickramasinghe. (August 30, 1990). "The Extragalactic Universe: An Alternative View" in *Nature*, vol. 346, issue 6287, 807–12 [doi:10.1038/346807a0].

Arp, Halton C. with Roy Keys and Konrad Rudnicki, editors. (1993). *Progress in New Cosmologies: Beyond the Big Bang*. New York: Plenum.

Arrhenius, Svante August. (1903). *Lehrbuch der Kosmischen Physik*. Leipzig: Verlag von S. Hirzel.

Arrhenius, Svante August. (1908). *Worlds in the Making: The Evolution of the Universe*, translation by H. Borns. New York: Harper.

Arrhenius, Svante August. (1909). *The Life of the Universe as Conceived by Man from the Earliest Ages to the Present Time*, translation by H. Borns. New York: Harper.

Arrhenius, Svante August. (1918). *The Destinies of the Stars*, translation by J. E. Fries. New York: Putnam.

Arrhenius, Svante August. (1925). *Chemistry in Modern Life*, translation and revision by C. S. Leonard. New York: D. Van Nostrand.

Artigas, Mariano. (2000). *The Mind of the Universe: Understanding Science and Religion*. Philadelphia: Templeton.

Ayala, Francisco J. (2008). "From Paley to Darwin: Design to Natural Selection" in *Back to Darwin: A Richer Account of Evolution*, edited by John Cobb. Grand Rapids: Eerdmans.

Babcock, William, translator. (2012). *The City of God—The Works of Saint Augustine: A Translation for the 21st Century*. New York City: New York City Press.

Badiou, Alain (2005). *Being and Event*, translation by Oliver Feltham. New York: Continuum.

Baker-Fletcher, Karen. (1998). *Sisters of Dust, Sisters of Spirit: Womanist Wordings on God and Creation*. Minneapolis: Fortress.

Baker-Fletcher, Karen. (2006). *Dancing with God: Trinity from a Womanist Perspective*. St. Louis: Chalice.

Bailey, M. E. with S. V. M. Clube and W. M. Napier. (1990). *The Origin of Comets*. Oxford: Pergamon.

Balashov, Yuri V. (1991). "Resource Letter AP-1: The Anthropic Principle" in *American Journal of Physics*, vol. 59, issue 12, 1069–76.

Balashov, Yuri V. (December 1994). "Uniformitarianism in Cosmology: Background and Philosophical Implications of the Steady-State Theory" in *Studies in History and Philosophy of Science Part A*, vol. 25, issue 6, 933–58.

Barad, Karen. (2007). *Meeting the Universe Halfway: Quantum Physics and the Entanglement of Matter and Meaning*. Durham: Duke University Press.

Barbour, Ian G. (1966). "Freedom as Indeterminacy" in *Issues in Science and Religion*. Englewood Cliffs: Prentice-Hall.

Barbour, Ian G. (1974). *Myths, Models, and Paradigms: A Comparative Study in Science and Religion*. New York: Harper.

Barbour, Ian G. (1990). *Religion in an Age of Science* (Gifford Lectures, Vol. I, 1989). San Francisco: Harper.

Barbour, Ian G. (1993). *Ethics in an Age of Technology* (Gifford Lectures, Vol. II, 1990). San Francisco: Harper.

Barbour, Ian G. (1997). *Religion and Science: Historical and Contemporary Issues* [revision of *Religion in an Age of Science* (1990)]. San Francisco: Harper.

Barbour, Ian G. (2000). *When Science Meets Religion: Enemies, Strangers, or Partners?* San Francisco: Harper.

Barbour, Ian G. (2008). "Evolution and Process Thought" in *Back to Darwin: A Richer Account of Evolution*, edited by John Cobb. Grand Rapids: Eerdmans.

Barker, Eileen. (1981). "Science as Theology—The Theological Functioning of Western Science" in *The Sciences and Theology in the Twentieth Century*, edited by A. R. Peacocke. Notre Dame: University of Notre Dame Press.

Barlow, Connie, editor. (1994). *Evolution Extended: Biological Debates on the Meaning of Life*. Cambridge: MIT Press.

Barnard, G. William (1997). *Exploring Unseen Worlds: William James and the Philosophy of Mysticism*. Albany: State University of New York Press.

Barnard, G. William (2011). *Living Consciousness: Reclaiming the Metaphysical Vision of Henri Bergson*. Albany: State University of New York Press.

Barrett, Justin L. (2011). *Cognitive Science, Religion, and Theology: From Human Minds to Divine Minds*. West Conshohocken: Templeton.

Barrow, John D. (2002). *The Constants of Nature: From Alpha to Omega, the Numbers that Encode the Deepest Secrets of the Universe*. New York: Pantheon.

Barrow, John D. (2005 [1995]). *The Artful Universe Expanded*. Oxford: Oxford University Press.

Barrow, John D. (2005). "Worlds without End or Beginning" in *The Scientific Legacy of Fred Hoyle*, edited by Douglas Gough, foreword by Martin Rees. Cambridge: Cambridge University Press.

Barrow, John D. (2007 [1991]). *New Theories of Everything: The Quest for Ultimate Explanation*. Oxford: Oxford University Press.

Barrow, John D. (2008). *Cosmic Imagery: Key Images in the History of Science*. New York: Norton.

Barrow, John D. (2011). *The Book of Universes: Exploring the Limits of the Cosmos*. New York: Norton.

Barrow, John D. and Michael S. Turner. (August 26, 1982). "Inflationary Universe—Birth, Death, and Transfiguration" in *Nature*, vol. 298, issue 5877, 801–5 [doi:10.1038/298801a0].

Barrow, John D. and Frank J. Tipler (1986). *The Anthropic Cosmological Principle*, foreword by John A. Wheeler. Oxford: Oxford University Press.

Barrow, John D. and Joseph Silk. (1993 [1983]). *The Left Hand of Creation: The Origin and Evolution of the Expanding Universe*. Oxford: Oxford University Press.

Barrow, John D. and John K. Webb. (February 1, 2012). "Inconstant Constants" in *Scientific American*, vol. 21, 70–77 [doi:10.1038/scientific-americantime0212–70].

Bedau, Mark A. and Carol E. Cleland, editors. (2010). *The Nature of Life: Classical and Contemporary Perspectives from Philosophy and Science*. Cambridge: Cambridge University Press.

Begbie, Jeremy S. (2013). "Natural Theology and Music" in *The Oxford Handbook of Natural Theology*, edited by Russell Re Manning. Oxford: Oxford University Press.

Behe, Michael J. (1996). *Darwin's Black Box: The Biochemical Challenge to Evolution*. New York: Free Press.

Bellantoni, Lisa. (2000). *Moral Progress: A Process Critique of MacIntyre*. Albany: State University of New York Press.

Bellantoni, Lisa (2004). "Future Ethics: MacIntyre and Whitehead on Moral Progress" in *Whitehead's Philosophy: Points of Connection*, edited by Janusz A. Polanowski and Donald W. Sherburne. Albany: State University of New York Press.

Ben-Jacob, Eshel. (June 7, 2014). "Do Bacteria Possess Social Intelligence?" in "The Forum" broadcast by BBC World Services, online at http://www.bbc.co.uk/programmes/p02042rf.

Bennett-Hunter, Guy. (2013). "Natural Theology and Literature" in *The Oxford Handbook of Natural Theology*, edited by Russell Re Manning. Oxford: Oxford University Press.

Benson, Michael. (2014). *Cosmigraphics: Picturing Space through Time.* New York: Abrams.

Berry, R. J. with D. C. Burke and others. (April 22, 1993). "Theology and Science" in *Nature*, vol. 362, 690 [doi:10.1038/362690b0].

Berry, R. J. (1996). *God and the Biologist: Faith at the Frontiers of Science.* Leicester: Apollos.

Bersanelli, Marco and Mario Gargantini. (2009). *From Galileo to Gell-Mann: The Wonder that Inspired the Greatest Scientists of All Time—In Their Own Words*, translation by J. Bowden. West Conshohocken: Templeton.

Bertka, Constance M., editor. (2009). *Exploring the Origin, Extent, and Future of Life: Philosophical, Ethical, and Theological Perspectives.* Cambridge: Cambridge University Press.

Bertotti, B. with R. Balbinot, S. Bergia, and A. Messina. (1990). *Modern Cosmology in Retrospect.* Cambridge: Cambridge University Press.

Binet, Alfred. (1889 [1888]). *The Psychic Life of Micro-Organisms: A Study in Experimental Psychology*, translation from French by T. McCormack. Chicago: Open Court.

Binet, Alfred. (1890). *On Double Consciousness: Experimental Psychological Studies.* Chicago: Open Court.

Binet, Alfred. (1907). *The Mind and the Brain.* London: K. Paul, Trench, Trubner.

Birch, Charles. (1965). *Nature and God.* Philadelphia: Westminster.

Birch, Charles. (1990). *A Purpose for Everything: Religion in a Postmodern Worldview.* Mystic: Twenty-Third Publications.

Birch, Charles. (1992). "The Postmodern Challenge to Biology" in *The Postmodern Reader*, edited by C. Jencks. New York: St. Martin's.

Birch, Charles. (2008). *Science and Soul.* Philadelphia: Templeton.

Birch, Charles and John B. Cobb Jr. (1981). *The Liberation of Life: From the Cell to the Community.* Cambridge: Cambridge University Press.

Bixler, J. S. (1941). "Whitehead's Philosophy of Religion" in *The Philosophy of Alfred North Whitehead, The Library of Living Philosophers, Vol. 3*, edited by P. A. Schilpp. La Salle: Open Court.

Blachowicz, James. (2012). *Essential Difference: Toward a Metaphysics of Emergence.* Albany: State University of New York Press.

Bockelée-Morvan, D. with D. Gautier and others. (May 1998). "Deuterated Water in Comet C/1996 B2 (Hyakutake) and its Implications for the Origin of Comets" in *Icarus*, vol. 133, issue 1, 147–62 [doi:10.1006/icar.1998.5916].

Bohm, David. (1992). "Postmodern Science and a Postmodern World" in *The Postmodern Reader*, edited by C. Jencks. New York: St. Martin's.

Bohren, Craig F. and N. C. Wickramasinghe. (August [16 February] 1977). "On the Computation of Optical Properties of Heterogeneous Grains" in *Astrophysics and Space Science*, vol. 50, issue 2, 461–72.

Boisvert, Kate Grayson. (2008). *Religion and the Physical Sciences.* Westport: Greenwood.

Bondi, Hermann. (August 5, 1947). "Spherically Symmetrical Models in General Relativity" in *Monthly Notices of the Royal Astronomical Society*, vol. 107, 410–25.

Bondi, Hermann. (1952). *Cosmology*. Cambridge: Cambridge University Press [1961 second edition; 2010 Dover edition].

Bondi, Hermann. (1960). *The Universe at Large*. Garden City: Anchor.

Bondi, Hermann. (1964). *Relativity and Common Sense: A New Approach to Einstein*. Garden City: Anchor.

Bondi, Hermann. (1966). "Some Philosophical Problems in Cosmology" in *British Philosophy in the Mid-Century*, edited by C. A. Mace. London: Allen and Unwin. (Kragh 1996: 304)

Bondi, Hermann. (2010). *Cosmology*, introduction by Ian Roxburgh. Mineola: Dover [previously Cambridge University Press 1952, 1961].

Bondi, Hermann and Fred Hoyle. (October 7, 1944). "On the Mechanism of Accretion by Stars" in *Monthly Notices of the Royal Astronomical Society*, vol. 104, issue 5, 21, 273–82.

Bondi, Hermann with Fred Hoyle and R. A. Lyttleton. (February 24, 1947). "On the Structure of the Solar Corona and Chromosphere" in *Monthly Notices of the Royal Astronomical Society*, vol. 107, issue 2, 184–210.

Bondi, Hermann and T. Gold. (July 14, 1948). "The Steady-State Theory of the Expanding Universe" in *Monthly Notices of the Royal Astronomical Society*, vol. 108, issue 3, 252–70.

Bondi, Hermann and Edwin E. Salpeter. (February 23, 1952). "Thermonuclear Reactions and Astrophysics" in *Nature*, vol. 169, issue 4295, 304–5 [doi:10.1038/169304a0].

Bondi, Hermann and others. (1960). *Rival Theories of Cosmology: A Symposium and Discussion of Modern Theories of the Structure of the Universe*. Oxford: Oxford University Press.

Bondi, Hermann and others. (1971). *Pioneering in Outer Space: A Course of Lectures on Selected Topics in Modern Physics and Space Flight*. London: Heinemann.

Bondi, Hermann with Thomas Gold and Fred Hoyle. (January 5, 1995). "Origins of Steady-State Theory" in *Nature*, vol. 373, issue 6509, 10 [doi:10.1038/373010b0].

Bondi, Hermann and Miranda Weston-Smith, editors. (1998). *The Universe Unfolding*. Oxford: Oxford University Press.

Bostrom, Nick. (2002). *Anthropic Bias: Observation Selection Effects in Science and Philosophy*, edited by R. Nozick. New York: Routledge.

Boto, Luis. (2010 [October 28, 2009]). "Horizontal Gene Transfer in Evolution: Facts and Challenges" in *Proceedings of the Royal Society B: Biological Sciences*, vol. 277, issue 1683, 819–27 [doi:10.1098/rspb.2009.1679].

Bould, Mark and Sherryl Vint. (2011). *The Routledge Concise History of Science Fiction*. New York: Routledge.

Bowker, John Westerdale. (1973). *The Sense of God; Sociological, Anthropological, and Psychological Approaches to the Origin of the Sense of God*. Oxford: Claredon.

Bowker, John Westerdale. (1978). *The Religious Imagination and the Sense of God.* Oxford: Oxford University Press; reprint 2000.

Bowker, John Westerdale. (1981). "Did God Create This Universe?" in *The Sciences and Theology in the Twentieth Century*, edited by A. R. Peacocke. Notre Dame: University of Notre Dame Press.

Bowker, John. (1987). *Licensed Insanities: Religion and Belief in God in the Contemporary World.* Oxford: Oxford University Press.

Bowker, John. (1995). *Is God a Virus?: Genes, Culture and Religion.* London: SPCK.

Bowker, John. (2005). *The Sacred Neuron: The Extraordinary New Discoveries Linking Science and Religion.* London: I. B. Tauris.

Bowker, John. (2014). *God: A Very Short Introduction.* Oxford: Oxford University Press.

Bowker, John, editor. (2008). *Knowing the Unknowable: Science and Religions on God and the Universe.* London: I. B. Tauris.

Bracken, Joseph A. (2004). "Panentheism: A Field-Oriented Approach" in *In Whom We Live and Move and Have Our Being: Panentheistic Reflections on God's Presence in a Scientific World*, edited by P. Clayton and A. Peacocke. Grand Rapids: Eerdmans.

Bracken, Joseph A. (2009). *Subjectivity, Objectivity, and Intersubjectivity: A New Paradigm for Religion and Science*, foreword by W. R. Stoeger. West Conshohocken: Templeton.

Bracken, Joseph A. (2012). *Does God Roll Dice? Divine Providence for a World in the Making.* Collegeville: Liturgical.

Bråkenhielm, Carl Reinhold, editor. (2009). *Linnaeus and Homo Religiosus: Biological Roots of Religious Awareness and Human Identity.* Proceedings from a conference at the Faculty of Theology, Uppsala University, May 30–June 2, 2007, in celebration of the birth of Carl von Linnaeus.

Brandt, John C. and Robert D. Chapman. (2004 [2002]). "Comets in History" in *Introduction to Comets.* Cambridge: Cambridge University Press.

Brierley, Michael W. (2004). "Naming a Quiet Revolution: The Panentheistic Turn in Modern Theology" in *In Whom We Live and Move and Have Our Being: Panentheistic Reflections on God's Presence in a Scientific World*, edited by P. Clayton and A. Peacocke. Grand Rapids: Eerdmans.

Broadnax, Reginald David. (2003). "The Category of Freedom in the Thought of Charles Hartshorne." PhD Dissertation, Northwestern University.

Brown, Cynthia Stokes. (2007). *Big History: From the Big Bang to the Present.* New York: New Press.

Brumbaugh, Robert S. (1982). *Whitehead, Process Philosophy, and Education.* Albany: State University of New York Press.

Brunner, Emil and Karl Barth. (2002 [1946]). *Natural Theology: Comprising "Nature and Grace" by Professor Dr. Emil Brunner and the reply "No!" by Dr. Karl Barth*, translation by P. Franenkel. Eugene: Wipf and Stock.

Brush, Stephen G. (August 1992). "How Cosmology Became a Science" in *Scientific American*, vol. 267, 62–70 [doi:10.1038/scientificamerican0892–62].

Burbidge, E. Margaret. (September 1994). "Watchers of the Skies" in *Annual Review of Astronomy and Astrophysics*, vol. 32, 1–37 [doi:10.1146/annurev.aa.32.090194.000245].

Burbidge, E. Margaret. (2005). "Modern Alchemy: Fred Hoyle and Element Building by Neutron Capture" in *The Scientific Legacy of Fred Hoyle*, edited by D. Gough, foreword by M. Rees. Cambridge: Cambridge University Press.

Burbidge, E. Margaret with Geoffrey R. Burbidge, William A. Fowler and Fred Hoyle. (October 1957). "Synthesis of the Elements in Stars" [informally identified by author initials as "B^2FH" (Rees 1997: 16), or as "B^2HF" (Kragh 1996: 300–4)] in *Review of Modern Physics*, vol. 29, issue 4, 547–650 [doi:10.1103/RevModPhys.29.547].

Burbidge, Geoffrey R. (June 17, 1961). "Galactic Explosions as Sources of Radio Emission" in *Nature*, vol. 190, issue 4781, 1053–56 [doi:10.1038/1901053a0].

Burbidge, Geoffrey. (August 1966). "The Origin of Cosmic Rays" in *Scientific American*, vol. 215, 32–38 [doi:10.1038/scientificamerican0866–32].

Burbidge, Geoffrey. (1967). *Lectures on High Energy Astrophysics*. Bombay: Tata Institute.

Burbidge, Geoffrey. (September 3, 1971). "Was there Really a Big Bang?" in *Nature*, vol. 233, issue 5314, 36–40 [doi:10.1038/233036a0].

Burbidge, G. R. (November 12, 1973). "Problem of the Redshifts" in *Nature Physical Science*, vol. 246, 17–25 [doi:10.1038/physci246017a0].

Burbidge, Geoffrey. (February 1992). "Why Only One Big Bang?" in *Scientific American*, vol. 266, 120 [doi:10.1038/scientificamerican0292–120].

Burbidge, Geoffrey. (2003). "Sir Fred Hoyle. 24 June 1915–20 August 2001, Elected FRS 1957" in *Biographical Memoirs of Fellows of the Royal Society*, vol. 49, 213–47 [doi:10.1098/rsbm.2003.0013].

Burbidge, Geoffrey. (2008a). "B^2FH, the Cosmic Microwave Background and Cosmology" in *Publications of the Astronomical Society of Australia*, vol. 25, issue 1, 30–35 [doi:10.1071/AS07029].

Burbidge, Geoffrey. (2008b). "Hoyle's Role in B^2FH" in *Science*, vol. 319, issue 5869, 1484b [doi:10.1126/science.319.5869.1484b].

Burbidge, Geoffrey. (November 14 [June 16], 2008). "A Realistic Cosmological Model Based on Observations and Some Theory Developed over the Last 90 Years" in *Italian Physical Society*, issue 12, 1437–52 [doi:10.1393/ncb/i2008–10486–9].

Burbidge, Geoffrey and Fred Hoyle. (December 1966). "The Problem of the Quasi-Stellar Objects" in *Scientific American*, vol. 215, 40–52 [doi:10.1038/scientificamerican1266–40].

Burbidge, Geoffrey and E. Margaret Burbidge. (1967). *Quasi-Stellar Objects*. San Francisco: Freeman.

Burbidge, Geoffrey R. and Fred Hoyle. (March 1, 1969). "Condensed Objects in the Crab Nebula" in *Nature*, vol. 221, issue 5183, 847–48 [doi:10.1038/221847a0].

Burbidge, Geoffrey R. and E. Margaret Burbidge. (October 4, 1969). "Quasi-stellar Objects—A Progress Report" in *Nature*, vol. 224, issue 5214, 21–24 [doi:10.1038/224021a0].

Burchell, M. J. with J. R. Mann and A. W. Bunch. (May 11, 2004). "Survival of Bacteria and Spores under Extreme Shock Pressures" in *Monthly Notices of the Royal Astronomical Society*, vol. 352, issue 4, 1273–78 [doi:10.1111/j.1365–2966.2004.08015.x].

Burke, Bernard and Francis Graham-Smith. (2002 [1997]). *An Introduction to Radio Astronomy*. Cambridge: Cambridge University Press.

Burns, Joseph A. (July 29, 2010). "Four Hundred Years of Planetary Science since Galileo and Kepler" in *Nature*, vol. 466, issue 7306, 575–84 [doi:10.1038/nature09215].

Burrell, David. (1973). *Analogy and Philosophical Language*. New Haven: Yale University Press.

Burrell, David B. with Carlo Cogliati, Janet M. Soskice, and William R. Stoeger, editors. (2010). *Creation and the God of Abraham*. Cambridge: Cambridge University Press.

Butler, Christopher. (2002). *Post-Modernism: A Very Short Introduction*. Oxford: Oxford University Press.

Butler, Declan. (September 7, 2006). "When Science and Theology Meet" in *Nature*, vol. 443, issue 7107, 10–11 [doi:10.1038/443010a].

Butler, Octavia E. (1993). *Parable of the Sower*. New York: Warner.

Butler, Octavia E. (1998). *Parable of the Talents*. New York: Seven Stories.

Cain, A. J. (April 24, 1980). "Theology after Darwin" [review of *Post-Darwinian Controversies* (1979) by James R. Moore] in *Nature*, vol. 284, issue 5758, 689–90 [doi:10.1038/284689a0].

Campbell, Joseph. (1985). *The Inner Reaches of Outer Space: Metaphor as Myth and as Religion*. New York: A van der Marck.

Campbell, Richard. (1990 [1976]). *From Belief to Understanding: A Study of Anselm's Proslogion Argument on the Existence of God*. Lewiston: Edwin Mellen.

Campbell, Ted. (1991a). *John Wesley and Christian Antiquity: Religious Vision and Cultural Change*. Nashville: Abingdon.

Campbell, Ted. (1991b). *The Religion of the Heart: A Study of European Religious Life in the Seventeenth and Eighteenth Centuries*. Columbia: University of South Carolina Press.

Campbell, Ted. (2010). *Wesleyan Beliefs: Formal and Popular Expressions of the Core Beliefs of Wesleyan Communities*. Nashville: Abingdon.

Campbell, Ted. (2011 [1999]). *Methodist Doctrine: The Essentials*. Nashville: Abingdon.

Campbell, Ted and Michael T. Burns (2004). *Wesleyan Essentials in a Multicultural Society*. Nashville: Abingdon.

Cannon, Katie Geneva. (1995). *Katie's Canon: Womanism and the Soul of the Black Community*. New York: Continuum.

Carr, Bernard J. (March 1982). "On the Origin, Evolution and Purpose of the Physical Universe" in *Irish Astronomical Journal*, vol. 15, issue 3, 237–53.

Carr, Bernard J. (2007). *Universe or Multiverse?* Cambridge: Cambridge University Press.

Carr, Bernard J. (April 2008). "Defending the Multiverse" in *Astronomy and Geophysics*, vol. 49, issue 2, 36–37 [doi:10.1111/j.1468–4004.2008.49229_2.x].

Carr, Bernard J. and Martin J. Rees. (April 12, 1979). "The Anthropic Principle and the Structure of the Physical World" in *Nature*, vol. 278, issue 5705, 605–12 [doi:10.1038/278605a0].

Carter, Brandon. (1974). "Large Number Coincidences and the Anthropic Principle in Cosmology" in *Confrontation of Cosmological Theories with Observational Data*, IAU Symposium, no. 63, 291–98, edited by M. S. Longair. Boston: D. Reidel.

Carter, Brandon. (1993). "The Anthropic Selection Principle and the Ultra-Darwinian Synthesis" in *Anthropic Principle: Proceedings of the Second Venice Conference on Cosmology and Philosophy*, edited by F. Bertola and U. Curi. Cambridge: Cambridge University Press.

Carter, Brandon. (March 1, 2004). "Anthropic Interpretation of Quantum Theory" in *International Journal of Theoretical Physics*, vol. 43, 721–30.

Carter, Brandon. (2007a). "Microanthropic Principle for Quantum Theory" in *Universe or Multiverse*, edited by B. Carr. Cambridge: Cambridge University Press.

Carter, Brandon. (October 18, 2007b). "The Significance of Numerical Coincidences in Nature" in *High Energy Physics—Theory* (Cornell University Library online version: ArXiv:0710.3543v1).

Carter, Brandon. (April–June 2011). "Classical Anthropic Everett Model: Indeterminacy in a Preordained Multiverse" in *Journal of Cosmology*, vol. 14, XIV [ArXiv.org/pdf/1203.0952].

Caruso, Gregg D., editor. (2014). *Science and Religion: Five Questions*. Copenhagen: Automatic.

Cauthen, Kenneth. (1984). *Process Ethics: A Constructive System*. Lewiston: Edwin Mellen.

Cauthen, Kenneth. (1991). *Theological Biology: The Case for a New Modernism*. Lewiston: Edwin Mellen.

Cauthen, Kenneth. (1997). *Toward a New Modernism*. Landham: University Press of America.

Cavanaugh, Michael A. (May 16, 1985). "Scientific Creationism and Rationality" in *Nature*, vol. 315, issue 6016, 185–89 [doi:10.1038/315185a0].

Ceuasco, George A. (May 1954). "Universe and Abbe Lemaître" in *Irish Monthly*, vol. 83, issue 969, 185–89.

Chaisson, Eric. (2006). *The Epic of Evolution: Seven Ages of the Cosmos*. New York: Columbia University Press.

Chalmers, David John. (1996). *Conscious Mind: In Search of a Fundamental Theory*. Oxford: Oxford University Press.

Chalmers, David John. (2002). *Philosophy of Mind: Classical and Contemporary Readings*. Oxford: Oxford University Press.

Chalmers, David John. (2010). *The Character of Consciousness*. Oxford: Oxford University Press.

Chalmers, David John. (2012). *Constructing the World*. Oxford: Oxford University Press.

Chen, Cheng-Yih. (1996). *Early Chinese Work in Natural Science: A Re-examination of the Physics of Motion, Acoustics, Astronomy and Scientific Thought*. Hong Kong: Hong Kong University Press.

Chiar, Jean E. and Yvonne J. Pendleton. (2008). "Origin and Evolution of Interstellar Organics" in *Organic Matter in Space: Proceedings of the 251st Symposium of the International Astronomical Union*, February 18–22, Hong Kong, China, edited by Sun Kwok and Scott A. Sandford. Cambridge: Cambridge University Press.

Chopra, Deepak. (March 2014). "Reality and Consciousness: A View from the East: Comment on 'Consciousness in the Universe: A Review of the 'Orch OR' Theory' by Hameroff and Penrose" in *Physics of Life Reviews*, vol. 11, issue 1, 81–82.

Chopra, Deepak and Leonard Mlodinow. (2011). *War of the Worldviews: Science vs Spirituality*. New York: Harmony.

Chyba, Christoper F. (December 23, 1987). "Cometary Contribution to the Oceans of Primitive Earth" in *Nature*, vol. 330, issue 6149, 632–35 [doi:10.1038/330632a0].

Chyba, Christoper and Carl Sagan. (September 17, 1987). "Cometary Organics but No Evidence for Bacteria" in *Nature*, vol. 329, issue 6136, 208 [doi:10.1038/329208a0].

Chyba, Christoper and Carl Sagan. (April 14, 1988). "Cometary Organic Matter Still a Contentious Issue" in *Nature*, vol. 332, issue 6165, 592 [doi:10.1038/332592a0].

Clarke, Arthur C. (2000 [1999]). "The Twenty-First Century: A (Very) Brief History" in *Greetings, Carbon-based Bipeds!: Collected Essays, 1934–1998*, edited by Ian T. Macauley. New York: St. Martin's.

Claus, George and Bartholomew Nagy. (November 18, 1961). "A Microbiological Examination of Some Carbonaceous Chondrites" in *Nature*, vol. 192, issue 4803, 594–96 [doi:10.1038/192594a0].

Clayton, Philip. (1989). *Explanation from Physics to Theology: An Essay in Rationality and Religion*. New Haven: Yale University Press.

Clayton, Philip. (1997). *God and Contemporary Science*. Grand Rapids: Eerdmans.

Clayton, Philip. (2000). *The Problem of God in Modern Thought*. Grand Rapids: Eerdmans.

Clayton, Philip. (2004a). "Panentheism in Metaphysical and Scientific Perspective" and "Panentheism Today: A Constructive Systematic Evaluation" in *In Whom We Live and Move and Have Our Being: Panentheistic Reflections on God's Presence in a Scientific World*, edited by P. Clayton and A. Peacocke. Grand Rapids: Eerdmans.

Clayton, Philip. (2004b). *Mind and Emergence: From Quantum to Consciousness*. Oxford: Oxford University Press.

Clayton, Philip. (2008). "Process and Emergence" in *Back to Darwin: A Richer Account of Evolution*, edited by John Cobb. Grand Rapids: Eerdmans.
Clayton, Philip. (2009). *In Quest of Freedom: The Emergence of Spirit in the Natural World*, edited by M. G. Parker and T. M. Schmidt. Golttingen: Vandenboeck & Ruprecht.
Clayton, Philip. (2012). *Religion and Science: The Basics*. New York: Routledge.
Clayton, Philip and Arthur Peacocke, editors. (2004). *In Whom We Live and Move and Have Our Being: Panentheistic Reflections on God's Presence in a Scientific World*. Grand Rapids: Eerdmans.
Clayton, Philip and Jeffrey Schloss, editors. (2004). *Evolution and Ethics: Human Morality in Biological and Religious Perspective*. Grand Rapids: Eerdmans.
Clayton, Philip and Sachary Simpson, editors. (2006). *The Oxford Handbook of Religion and Science*. Oxford: Oxford University Press.
Cobb, John B. Jr. (Spring 1964). "From Crisis Theology to the Post-Modern World" in *Centennial Review*, vol. 8, 209–20.
Cobb, John B. Jr. (1988). "Ecology, Science, and Religion: Toward a Postmodern Worldview" in *The Reenchantment of Science: Postmodern Proposals*, edited by David Ray Griffin. Albany: State University of New York Press.
Cobb, John B. Jr. (2004). "Thinking with Whitehead about Nature" in *Whitehead's Philosophy: Points of Connection*, edited by Janusz A. Polanowski and Donald W. Sherburne. Albany: State University of New York Press.
Cobb, John B. Jr. (2008). "Organisms as Agents in Evolution" in *Back to Darwin: A Richer Account of Evolution*, edited by J. Cobb. Grand Rapids: Eerdmans.
Cobb, John B. Jr. and W. Widick Schroeder, editors. (1981). *Process Philosophy and Social Thought*. Chicago: Center for the Scientific Study of Religion.
Cobb, John B. Jr. and Franklin I. Gamwell. (1984). *Existence and Actuality: Conversations with Charles Hartshorne*. Chicago: University of Chicago Press.
Cole, Shaun with Cederic G. Lacey, Carlton M. Baugh, and Carlos S. Frenk. (November [4 April] 2000). "Hierarchical Galaxy Formation" in *Monthly Notices of the Royal Astronomical Society*, vol. 319, issue 1, 168–204 [doi:10.1046/j.1365–8711.2000.03879.x].
Cole, Shaun with Peder Norberg, Carlton M. Baugh, Carlos S. Frenk, and others. (September 2001). "The 2dF Galaxy Redshift Survey: Near-Infrared Galaxy Luminosity Functions" in *Monthly Notices of the Royal Astronomical Society*, vol. 326, issue 1, 255–73 [doi:10.1046/j.1365–8711.2001.04591.x].
Coleman, Monica A. (2008). *Making a Way Out of No Way: A Womanist Theology*. Minneapolis: Fortress.
Coleman, Monica A. with Nancy R. Howell and Helene Tallon Russell, editors. (2011). *Creating Women's Theology: A Movement Engaging Process Thought*. Eugene: Pickwick.
Coleman, Monica A. (2013). *Ain't I A Womanist, Too?: Third Wave Womanist Religious Thought*. Minneapolis: Fortress.
Coleridge, Samuel Taylor. (1798). "The Rime of the Ancient Mariner" in *Lyrical Ballads* (Bristol: Printed by Biggs and Cottle for T. N. Longman; London:

Printed for J. and A. Arch, 1798) by William Wordsworth and Samuel Taylor Coleridge [reprinted in *Lyrical Ballads 1798: A Critical Edition*, with introduction and notes by E. Dowden, edited by W. K. Chapman. Clemson University Digital Press, 2011 [www.clemson.edu/cedp/cudp/pubs/ballads/ballads.pdf].

Coleridge, Samuel Taylor. (1848). *Hints towards the Formation of a More Comprehensive Theory of Life*, edited by Seth B. Watson. M. D. London: John Churchill; Farnborough: Gregg International.

Coleridge, Samuel Taylor. (1859 [1854]). *A Dissertation on the Science of Method; Or, The Laws and Regulative Principles of Education*, eighth edition. London: Charles Griffin.

Coles, Peter, editor. (1998). *Icon Critical Dictionary of the New Cosmology*, foreword by John D. Barrow. Cambridge: Icon.

Coles, Peter, editor. (1999). *The Routledge Critical Dictionary of the New Cosmology.* New York: Routledge.

Coles, Peter. (2001). *Cosmology: A Very Short Introduction.* Oxford: Oxford University Press.

Coles, Peter and George Ellis. (1997). *Is the Universe Open or Closed?* Cambridge: Cambridge University Press.

Coles, Peter and Grancesco Lucchin. (2002 [1995]). *Cosmology: The Origin and Evolution of Cosmic Structure.* Chichester: John Wiley.

Collins, Francis S. (2006). *The Language of God: A Scientist Presents Evidence for Belief.* New York: Free Press.

Collins, Kenneth J. (2011). "Today and Tomorrow: Recent Trends in Cosmology" in *The Theology of John Wesley: Holy Love and the Shape of Grace.* Nashville: Abingdon.

Collins, Kenneth J. and Jason E. Vickers, editors. (2013) *The Sermons of John Wesley: A Collection for the Christian Journey.* Nashville: Abingdon.

Consolmagno, Guy. (2000). *Brother Astronomer: Adventures of a Vatican Scientist.* New York: McGraw Hill.

Consolmagno, Guy. (2009). *The Heavens Proclaim: Astronomy and the Vatican.* Vatican City State: Vatican Observatory Publications.

Consolmagno, Guy. (Spring 2011). "Cosmic Origins" in *Cosmic Origins Lecture Series* at University of Arizona [YouTube title "Cosmology: Making Sense of the Universe"].

Cook, C. W. with W. A. Fowler, C. C. Lauritsen, and T. Lauritsen. (July 15, 1957). "B^{12}, C^{12}, and the Red Giants" in *Physical Review Letters*, vol. 107, issue 2, 508–15 [doi:10.1103/PhysRev.107.508].

Cooper, John W. (2006). *Panentheism—The Other God of the Philosophers: From Plato to the Present.* Grand Rapids: Baker.

Copernicus, Nicolaus. (1543). *De revolutionibus orbium coelestium, libri VI* ["On the Revolution of the Celestial Spheres, Book 6"]. Budapest: Tancsics Könyvkiado.

Corey, Michael Anthony. (1993a). *Back to Darwin: The Scientific Case for Deistic Evolution.* Lanham: Rowman and Littlefield.

Corey, Michael Anthony. (1993b). *God and the New Cosmology: The Anthropic Design Argument.* Lanham: Rowman and Littlefield.

Corey, Michael Anthony. (1995). *The Natural History of Creation: Biblical Evolutionism and the Return of Natural Theology.* Lanham: Rowman and Littlefield.

Cornwell, John. (2007). *Darwin's Angel: A Seraphic Response to the God Delusion.* London: Profile.

Coulson, S. G. and N. C. Wickramasinghe. (August 12, 2003). "Frictional and Radiation Heating of Micro-Sized Meteoroids in the Earth's Upper Atmosphere" in *Monthly Notices of the Royal Astronomical Society*, vol. 343, issue 4, 1123–30 [doi:10.1046/j.1365–8711.s003.06478.x].

Cowen, Ron. (September 27, 2013). "How to See Quantum Gravity in Big Bang Traces: Gravitons Could be Detectable in the Cosmic Microwave Background" in *Nature News* [doi:10.1038/nature.2013.13834].

Cowen, Ron. (March 17, 2014a). "How Astronomers Saw Gravitational Waves from the Big Bang" in *Nature News* [doi:10.1038/nature.2014.14885].

Cowen, Ron. (March 20, 2014b). "Telescope Captures View of Gravitational Waves: Images of the Infant Universe Reveal Evidence for Rapid Inflation after the Big Bang" in *Nature*, vol. 507, issue 7492, 281–83 [doi:10.1038/507281a].

Craig, William Lane. (1979). *The Kalam Cosmological Argument.* New York: Barnes and Noble.

Craig, William Lane. (2001). *The Cosmological Argument from Plato to Leibnitz.* Eugene: Wipf and Stock.

Craig, William Lane and Quentin Smith. (1993). *Theism, Atheism, and Big Bang Cosmology.* Oxford: Oxford University Press.

Craig, William Lane and Chad Meister, editors. (2009). *God Is Great, God Is Good: Why Believing in God Is Reasonable and Responsible.* Downers Grove: IVP Books.

Craig, William Lane and J. P. Moreland, editors. (2009). *The Blackwell Companion to Natural Theology.* Malden: Wiley-Blackwell.

Crawford, David F. (February–March 2011). "Observational Evidence Favors a Static Universe" in *Journal of Cosmology*, vol. 13, issue 20–22, 3875–4057 [arxiv.org/pdf/1009.0953.pdf].

Crean, Thomas. (2007). *God Is No Delusion: A Refutation of Richard Dawkins.* San Francisco: Ignatius.

Crease, Robert P. (February 19, 2015). "History of Science: Unshadowed Lens on the Past" [review of Steven Weinberg's *To Explain the World: The Discovery of Modern Science* (2015)] in *Nature*, vol. 518, issue 300 [doi:10.1038/518300a].

Crease, Robert P. and Charles C. Mann. (1996 [1986]). *The Second Creation: Makers of the Revolution in Twentieth-Century Physics*, foreword by Timothy Ferris. New Brunswick: Rutgers University Press.

Crick, F. H. C. (August 21, 1968). "The Origin of the Genetic Code" in *Journal of Molecular Biology*, vol. 38, 367–79.

Crick, Francis with Harry Compton and Leslie E. Orgel. (July 1973). "Directed Panspermia" in *Icarus*, vol. 19, issue 3, 341–46.

Crockett, Clayton. (2013). "Postmodernism and Natural Theology" in *The Oxford Handbook of Natural Theology*, edited by Russell Re Manning. Oxford: Oxford University Press.

Crutcher, Timothy. (2009). "The Consonance of Wesleyan Theology and Modern Science" in *Divine Grace and Emerging Creation: Wesleyan Forays in Science and Theology of Creation*, edited by Thomas Jay Oord. Eugene: Pickwick.

Cunningham, Mary Kathleen, editor. (2007). *God and Evolution: A Reader*. New York: Routledge.

Darling, David. (2001). *Life Everywhere: The Maverick Science of Astrobiology*. New York: Basic.

Darwin, Charles. (1859). *On the Origin of Species by Means of Natural Section, or the Preservation of Favoured Races in the Struggle for Life*. London: John Murray.

Das, Pranab Kumar. (2009). *Global Perspectives on Science and Spirituality*. West Conshohocken: Templeton.

Davaney, Sheila Greeve. (May 1986). *Divine Power: A Study of Karl Barth and Charles Hartshorne*. Harvard Dissertations in Religion.

Davidsson, Björn J. R. with Samuel Gulkis, Claudia Alexander, and others. (November 2010). "Gas Kinetics and Dust Dynamics in Low-Density Comet Comae" in *Icarus*, vol. 210, issue 1, 455–71 [doi:10.1016/j.icarus.2010.06.022].

Davies, Paul C. W. (1982). *Accidental Universe*. Cambridge: Cambridge University Press.

Davies, Paul C. W. (2004 [1998]). *The Cosmic Blueprint—New Discoveries in Nature's Creative Ability to Order the Universe*. Philadelphia: Templeton.

Davies, Paul C. W. (2004). "Teleology without Teleology: Purpose through Emergent Complexity" in *In Whom We Live and Move and Have Our Being: Panentheistic Reflections on God's Presence in a Scientific World*, edited by P. Clayton and A. Peacocke. Grand Rapids: Eerdmans.

Davies, Paul C. W. (2008). *The Goldilocks Enigma: Why Is the Universe Just Right for Life?* Boston: Mariner [originally *Cosmic Jackpot* (2006)].

Davies, Paul and Niels Henrik Gregersen, editors. (2010). *Information and the Nature of Reality: From Physics to Metaphysics*. Cambridge: Cambridge University Press.

Davis, Jimmy H. and Harry L. Poe. (2008). *Chance or Dance: An Evaluation of Design*. West Conshohocken: Templeton.

Dawkins, Richard. (1976). *The Selfish Gene*. Oxford: Oxford University Press; reprints 1989, 2006.

Dawkins, Richard. (1982). *Extended Phenotype: The Gene as the Unit of Selection*. San Francisco: Freeman.

Dawkins, Richard. (1983). *The Extended Phenotype: The Long Reach of the Gene*. Oxford: Oxford University Press; revised 1999.

Dawkins, Richard. (1987 [1986]). *The Blind Watchmaker: Why the Evidence of Evolution Reveals a Universe without Design*. New York: Norton; reprint 1994, reprint with new introduction 1996.

Dawkins, Richard. (1995). *Rivers Out of Eden: A Darwinian View of Life*. New York: Basic.

Dawkins, Richard. (1996). *Climbing Mount Improbable*, drawings by L. Ward. New York: Norton.

Dawkins, Richard. (1998). *Unweaving the Rainbow: Science, Delusion, and the Appetite for Wonder*. Boston: Houghton Mifflin.

Dawkins, Richard. (2003). *A Devil's Chaplain: Reflections on Hope, Lies, Science, and Love*. Boston: Houghton Mifflin.

Dawkins, Richard. (2004). *The Ancestor's Tale: A Pilgrimage to the Dawn of Evolution*. Boston: Houghton Mifflin.

Dawkins, Richard. (2006a). *The God Delusion*. Boston: Houghton Mifflin.

Dawkins, Richard. (2006b). *The Selfish Gene*, thirtieth anniversary edition. Oxford: Oxford University Press.

Dawkins, Richard. (2009). *The Greatest Show on Earth: The Evidence for Evolution*. New York: Free Press.

Dawkins, Richard. (2011). *The Magic of Reality: How We Know What's Really True*. New York: Free Press.

Dawkins, Richard. (2013). *An Appetite for Wonder: The Making of a Scientist: A Memoir*. London: Bantam.

Dawkins, Richard, editor. (2008). *The Oxford Book of Modern Science Writing*. Oxford: Oxford University Press.

Day, Vox. (2008). *The Irrational Atheist: Dissecting the Unholy Trinity of Dawkins, Harris, and Hitchens*. Dallas: Ben Bella.

Deamer, David W. (2012). *First Life: Discovering the Connections between Stars, Cells, and How Life Began*. Berkeley: University of California Press.

Dean, Cornelia, editor (2013). *The New York Times Book of Physics and Astronomy: More Than 100 Years of Covering the Expanding Universe*, foreword by Neil de Grasse Tyson. New York: Sterling.

Deane-Drummond, Celia. (1997). *Theology and Biotechnology: Implications for a New Science*. London: Geoffrey Chapman.

Deane-Drummond, Celia. (2000). *Creation through Wisdom: Theology and the New Biology*. Edinburgh: T&T Clark.

Deane-Drummond, Celia. (2003). *Brave New World?: Theology, Ethics, and the Human Genome*, foreword by Ted Peters. New York: T&T Clark.

Deane-Drummond, Celia. (2004). *The Ethics of Nature*. Malden: Wiley-Blackwell.

Deane-Drummond, Celia. (2006). *Wonder and Wisdom: Conversations in Science, Spirituality, and Theology*. Philadelphia: Templeton.

Deane-Drummond, Celia. (2008). *Eco-theology*. London: Darton, Longman and Todd.

Deane-Drummond, Celia with Bronislaw Szerzynske and Robin Grove-White, editors. (2003). *Re-ordering Nature: Theology, Society, and the New Genetics*. New York: T&T Clark.

Del Re, Giuseppe. (2000). *The Cosmic Dance: Science Discovers the Mysterious Harmony of the Universe*, foreword by Thomas Torrance. Philadelphia: Templeton.

Delio, Ilia. (2013). *The Unbearable Wholeness of Being: God, Evolution, and the Power of Love*. Maryknoll: Orbis.

Dellinger, Drew. (2011 [2002]). *Love Letter to the Milky Way: A Book of Poems.* Ashland: White Cloud Press.

Deloria, Vine Jr. (1979). "Theologians and Scientists" and "Transformation of Science" in *Metaphysics of Modern Existence.* San Francisco: Harper and Row.

Depoortere, Frederick. (2009). *Badiou and Theology.* New York: T & T Clark.

Derham, William. (1713). *Physico-Theology: Or, A Demonstration of the Being and Attributes of God from His Works of Creation.* London: William Innys.

Derham, William. (1715). *Astro-Theology: Or, A Demonstration of the Being and Attributes of God, from a Survey of the Heavens.* London: William Innys.

Devenish, Philip E. (October 4, 1981). "Mind, Brain, and Dualism" in *Journal of Religion*, vol. 61, issue 4, 422–27.

Devenish, Philip E. and George L. Goodwin, editors. (1989). *Witness and Existence: Essays in Honor of Schubert M. Ogden.* Chicago: University of Chicago Press.

Dewey, John (1941). "The Philosophy of Whitehead" in *The Philosophy of Alfred North Whitehead, The Library of Living Philosophers, Vol. 3,* edited by Paul Arthur Schilpp. La Salle: Open Court.

Dick, Steven J. (1993). "Plurality of Worlds" in *Cosmology: Historical, Literary, Philosophical, Religious, and Scientific Perspectives,* edited by N. S. Hetherington. New York: Garland; Routledge, 2008.

Dick, Steven J. (1996). *The Biological Universe: The Twentieth-Century Extraterrestrial Life Debate and the Limits of Science.* Cambridge: Cambridge University Press.

Dick, Steven J. (1998). *Life on Other Worlds: The Twentieth-Century Extraterrestrial Life Debate.* Cambridge: Cambridge University Press.

Dick, Steven J., editor. (2000). *Many Worlds: The New Universe, Extraterrestrial Life, and the Theological Implications.* Philadelphia: Templeton.

Dick, Steven J. and James E. Strick. (2005). *The Living Universe: NASA and the Development of Astrobiology.* New Brunswick: Rutgers University Press.

Dicke, R. H. (November 4, 1961). "Dirac's Cosmology and Mach's Principle" in *Nature,* vol. 192, issue 4801, 440–41 [doi:10.192440a0].

Dirac, P. A. M. (November 4, 1961). "Dirac's Cosmology and Mach's Principle" in *Nature,* vol. 192, issue 4801, 441 [doi:10.192441a0].

Dobretsov, Nikolay and others, editors. (2008). *Biosphere Origin and Evolution.* New York: Springer.

Dodelson, Scott. (2003). *Modern Cosmology.* San Diego: Academic.

Dombrowski, Daniel A. (2004). *Divine Beauty: The Aesthetics of Charles Hartshorne.* Vanderbilt University Press.

Donn, Bertram with N. C. Wickramasinghe, J. P. Hudson, and T. P. Stecher. (August 1968 [September 21, 1966]). "On the Formation of Graphite Grains in Cool Stars" in *The Astrophysical Journal,* vol. 153, 451–64 [doi:10.1086/149679].

Drees, William B. (1990). *Beyond the Big Bang: Quantum Cosmologies and God.* La Salle: Open Court.

Drees, William B. (June 1991). "Informed Intellect and Integrity" [concerning Ernan McMullin] in *Zygon: Journal of Religion and Science*, vol. 46, issue 2, 261–64 [doi:10.1111/j.1467–9744.2011.01194.x].

DuBay, Thomas. (1999). *The Evidential Power of Beauty: Science and Theology Meet*. San Francisco: Ignatius.

Duley, W. W. and D. A. Williams. (January 4, 1979). "Are There Organic Grains in the Interstellar Medium?" in *Nature*, vol. 277, issue 5691, 40–41 [doi:10.1038/277040a0].

Dunbar, D. N. F. with R. E. Pixley, W. A. Wenzel, and W. Whaling. (October 31, 1953). "The 7.68-MeV State in C^{12}" in *Physical Review*, vol. 92, issue 3, 649–50 [doi:10.1103/PhysRev.92.649].

Duncan, Todd and Craig Tyler. (2009). *Your Cosmic Context: An Introduction to Modern Cosmology*. San Francisco: Addison-Wesley.

de Duve, Christian. (c1995). *Vital Dust: Life as Cosmic Imperative*. New York: Basic Books.

Dyson, Freeman. (1988). *Infinite in All Directions* (Gifford Lectures, 1983–85 at Aberdeen) New York: Harper.

Dyson, Freeman. (May 27, 2015). "Stardust" [review of *Expanding Universe* (2015) by O. Edwards and Z. Levay] in the *New York Review of Books*.

Eagleton, Terry. (2009). *Reason, Faith and Revolution: Reflections on the God Debate*. New Haven: Yale University Press.

Eastman, Timothy E. (September 25, 2006). "Cosmic Agnosticism" paper presented at a conference on "Cosmology and Process Philosophy in Dialogue," October 5–8, Claremont School of Theology.

Eastman, Timothy E. (December 26, 2009). "Cosmic Agnosticism Revisited" in *Journey of Cosmology*, vol. 4, issue 2, 655–63.

Eccles, John C. (1979). *The Human Mystery* (1979 Gifford). Berlin: Springer-Verlag.

Eccles, John C. (1980). *The Human Psyche* (1980 Gifford). Berlin: Springer-Verlag.

Eccles, John C. (August 1992). "Evolution of Consciousness" in *Proceedings of the National Academy of Sciences*, vol. 89, 7320–24.

Edwards, Owen and Zoltan Levay. (2015). *Expanding Universe: Photographs from the Hubble Space Telescope*. New York: Taschen.

Edwards, Rem B. (Fall 1975). "The Human Self: An Actual Entity or a Society?" in *Process Studies*, vol. 5, issue 3, 195–203.

Edwards, Rem B. (1992). "God and Process" in *Logic, God, and Metaphysics*, edited by J. Harris. Dordrecht: Kluwer.

Edwards, Rem B. (Spring–Summer 2000). "How Process Theology Can Affirm Creation Ex Nihilo" in *Process Studies*, vol. 20, issue 1, 77–96.

Edwards, Rem B. (2001). *What Caused the Big Bang?* Amsterdam: Rodopi.

Edwards, Rem B. (2010). *The Essentials of Formal Axiology*. Lanham: University Press of America.

Edwards, Rem B. (2012). *John Wesley's Values and Ours*. Lexington: Emeth.

Edwards, Rem B. (Spring/Summer 2013). "God as a Single Processing Actual Entity" in *Process Studies*, vol. 42, issue 1, 77–86.

Efstathiou, George. (15 May 1992). "Suppressing the Formation of Dwarf Galaxies via Photoionization" in *Monthly Notices of the Royal Astronomical Society*, vol. 256, issue 2, 43P–47P.

Efstathiou, George and Bernard J. T. Jones. (1979 [June 21, 1978; March 13, 1978]). "Rotation of Galaxies: Numerical Investigations of the Tidal Torque Theory" in *Monthly Notices of the Royal Astronomical Society*, vol. 186, 133–44.

Ehrenfreund, Pascale and others. (2004). *Astrobiology: Future Perspectives*. Boston: Kluwer.

Eicher, David J. (2013). *Comets!: Visitors from Deep Space*. Cambridge: Cambridge University Press.

Einhorn, Stefan. (2002). *A Concealed God: Religion, Science, and the Search for Truth*, translation by L. Schenck. Philadelphia: Templeton.

Einstein, Albert. (1917). "Kosmologische Betrachtungen zur allgemeinen Relativitätstheorie" in *Sitzungsberichte der Königlich Preussischen Akademie der Wissenschaften*. Berlin.

Einstein, Albert. (1920 [December 1916]). *Relativity: The Special and General Theory*, translation by R. W. Lawson. London: Methuen; New York: H. Holt, 1920; Routledge, 2001.

Eliade, Mircea. (1974 [original French 1949]). *The Myth of the Eternal Return or, Cosmos and History*, 1954 translation by W. Trask. Princeton: Princeton University Press.

Elíasdóttir, Á. with J. P. U. Fynbo and others. (2009). "Dust Extinction in High-z Galaxies with Gamma-Ray Burst Afterglow Spectroscopy: The 2175 Å Feature at z = 2.45" in *Astrophysical Journal*, vol. 697, issue 2, 1725–40.

Ellis, George Francis Rayner. (1990). "Innovation, Resistance and Change: The Transition to the Expanding Universe" in *Modern Cosmology in Retrospect*, edited by B. Bertotti, R. Balbinot, S. Bergia, and A. Messina. Cambridge: Cambridge University Press.

Ellis, George F. R. (1993). *Before the Beginning: Cosmology Explained*. London: Boyars/Bowerdean.

Ellis, George F. R. (January 12–15, 2004). "Cosmology." Lectures 1–4 at the annual African Summer Theory Institute in Cape Town, South Africa, online at www.asti.ac.za/lectures.php.

Ellis, George F. R. (April 2008). "Opposing the Multiverse" in *Astronomy and Geophysics*, vol. 49, issue 2, 33–35 [doi:10.1111/j.1468–4004.2008.49229_1.x].

Ellis, George F. R. (January 20, 2011). "Cosmology: The Untestable Multiverse" in *Nature*, vol. 469, issue 7330, 294–95 [doi:10.1038/469294a].

Ellis, George F. R. (July 19, 2011). "Does the Multiverse Really Exist?" in *Scientific American*, vol. 305, 38–43 [doi:10.1038/scientificamerican0811–38].

Ellis, George F. R., editor. (2002). *Far-Future Universe: Eschatology from a Cosmic Perspective*. Philadelphia: Templeton.

Ellis, George F. R. with Roy Maartens and Malcolm A. H. MacCallum. (2012). *Relativistic Cosmology*. Cambridge: Cambridge University Press.

Ellis, George and Joe Silk. (December 16, 2014). "Defend the Integrity of Physics: Attempts to Exempt Speculative Theories of the Universe from Experimental Verification Undermine Science" in *Nature*, vol. 516, issue 7531, 321–23 [doi:10.1038/516321a].

Ellis, Richard S. with Ross J. McLure and others. (2012). "Abundance of Star-Forming Galaxies in the Redshift Range 8.5–12: New Results from the 2012 Hubble Ultra Deep Field Campaign" in *Astrophysical Journal Letters*, vol. 763, issue 1, L7 [arXiv:1211.6804v1].

Elsdon-Baker, Fern. (2009). *The Selfish Genius: How Richard Dawkins Rewrote Darwin's Legacy*. London: Icon.

Engel, Michael H. and Bartholomew Nagy. (April 29, 1982). "Distribution and Enantiomeric Composition of Amino Acids in the Murchinson Meteorite" in *Nature*, vol. 296, issue 5860, 837–40 [doi:10.1038/296837a0].

Engel, Michael H. and Stephen A. Macho. (September 18, 1997 [July 14, 1997; September 17, 1996]). "Isotropic Evidence for Extraterrestrial Non-Racemic Amino Acids in the Murchison Meteorite" in *Nature*, vol. 389, issue 6648, 265–68 [doi:10.1038/38460].

Engel, Michael H. with Stephen A. Macko and J. A. Silfer. (November 1, 1990). "Carbon Isotope Composition of Individual Amino Acids in the Murchison Meteorite" in *Nature*, vol. 348, issue 6296, 47–49 [doi:10.1038/34847a0].

Epperly, Bruce G. (2011). *Process Theology: A Guide to the Perplexed*. New York: T & T Clark.

Epperly, Bruce G. (2014). *Process Theology: Embracing Adventure with God*. Gonzalez: Energion.

Epperson, Michael. (2004). *Quantum Mechanics and the Philosophy of Alfred North Whitehead*. Bronx: Fordham University Press.

Epperson, Michael. (2013). *Foundations of Relational Realism: A Topical Approach to Quantum Mechanics and the Philosophy of Nature*. Lanham: Lexington Books.

Esan, Joe. (2009). *The Godless Delusion: Dawkins and the Limits of Human Sight*. New York: Peter Lang.

Evans, Sir Richard J. (March 14, 2011). "The Victorians: Religion and Science" (Gresham Lecture). London: Gresham College.

Ewart, Paul. (2013). "Physical Sciences and Natural Theology" in *The Oxford Handbook of Natural Theology*, edited by Russell Re Manning. Oxford: Oxford University Press.

Faber, Roland. (2008). *God as Poet of the World: Exploring Process Theologies*, translation by D. Scott. Louisville: Westminster John Knox.

Faber, Roland. (Spring/Summer 2012). "Three Hundred Years of Whitehead: Halfway" in *Process Studies*, vol. 41, issue 1, 5–20.

Fall, S. Michael and George Efstathiou. (March 17, 1980 [December 14, 1979]). "Formation and Rotation of Disc Galaxies with Haloes" in *Monthly Notices of the Royal Astronomical Society*, vol. 193, 189–206.

Faux, Michael and S. James Gates Jr. (March 9, 2005 [October 18, 2004]). "Adinkras: A Graphical Technology for Supersymmetric Representation Theory" in *Physical Review D*, vol. 71, issue 6 [doi:10.1103/PhysRevD.71.065002].

Felleman, Laura Bartels. (2009). "Degrees of Certainty in John Wesley's Natural Philosophy" in *Divine Grace and Emerging Creation: Wesleyan Forays in Science and Theology of Creation*, edited by Thomas Jay Oord. Eugene: Pickwick.

Fenn, Richard K. (2013). "A Sociological Perspective on Natural Theology" in *The Oxford Handbook of Natural Theology*, edited by Russell Re Manning. Oxford: Oxford University Press.

Ferguson, Kitty. (1995). *Fire in the Equations: Science, Religion, and the Search for God*. Grand Rapids: Eerdmans.

Fernández, Julio Angel. (2005). *Comets—Nature, Dynamics, Origin and Their Cosmological Relevance*. Dordrecht: Springer.

Ferré, Frederick. (1996). *Being and Value: Toward a Constructive Postmodern Metaphysics*. Albany: State University of New York Press.

Ferris, Timothy. (1997). *The Whole Shebang: A State of the Universe(s) Report*. New York: Touchstone.

Fitch, Frank with Henry P. Schwarcz and Edward Anders. (March 24, 1962). "'Organized Elements' in Carbonaceous Chondrites" in *Nature*, vol. 193, issue 4821, 1123–25 [doi:10.1038/1931123a0].

Fleisher, Paul. (2006). *The Big Bang (Great Ideas of Science)*. Minneapolis: Twenty-First Century.

Flew, Antony. (1966). *God and Philosophy*. New York: Harcourt.

Flew, Antony. (1967). *Evolutionary Ethics*. New York: St. Martin's.

Flew, Antony. (1976). *The Presumption of Atheism, and Other Philosophical Essays on God, Freedom, and Immortality*. New York: Barnes & Noble.

Flew, Antony. (1984). *God: A Critical Enquiry*. LaSalle: Open Court.

Flew, Antony. (1993). *Atheistic Humanism*. Buffalo: Prometheus.

Flew, Antony with Roy Abraham Varghese. (2007). *There Is a God: How the World's Most Notorious Atheist Changed His Mind / Antony Flew*. New York: Harper One.

Floyd-Thomas, Stacey M. and Miguel A. De La Torre, editors. (2011). *Beyond the Pale: Reading Ethics from the Margins*. Louisville: Westminster John Knox.

Folsome, Clair Edwin. (1979). *The Origin of Life: A Warm Little Pond*. San Francisco: Freeman.

Forterre, Patrick and Herve Philippe. (October 1999 [September 23]). "Where Is the Root of the Universal Tree of Life?" in *BioEssays*, vol. 21, issue 10, 871–79 [doi:10.1007/s11084–011–9254–1].

Fowler, William A. (1954). "Experimental and Theoretical Results on Nuclear Reactions in Stars" in *Mémoires de la Societé Royale des Sciences de Liegè*, vol. 13, 88–112.

Fowler, William A. (September 1956). "The Origin of the Elements" in *Scientific American*, vol. 195, 82–91 [doi:10.1038/scientificamerican0956–82].

Fowler, William A. (July 1965). *Neutrino Astrophysics: Supermassive Stars, Quasars, and Extragalactic Radio Sources and Nuclear Energy Generation in Supermassive Stars*. Kellogg Foundation: Orange Aid Preprint Series in Nuclear Astrophysics.

Fowler, William A. (1967). *Nuclear Astrophysics*. Philadelphia: American Philosophical Society.

Fowler, William A. (November 1984). "Quest for the Origin of the Elements" in *Science*, vol. 226, issue 4677, 922–35 [doi:10.1126/science.226.4677.922].

Fowler, William A. and Jesse L. Greenstein. (April 15, 1956). "Element-Building Reactions in Stars" in *Proceedings of the National Academy of Sciences*, vol. 42, issue 4, 173–80.

Fowler, William A. and Fred Hoyle. (1960). "Nucleosynthesis in Supernovae" in *The Astronomical Journal*, vol. 65, 345.

Fowler, William A. and Fred Hoyle. (November 1960 [May 21]). "Nucleosynthesis in Supernovae" in *Astrophysical Journal*, vol. 132, 565–90 [doi:10.1086/146963].

Fowler, William A. and Fred Hoyle. (December 1964 [April 8]). "Neutrino Processes and Pair Formation in Massive Stars and Supernovae" in *The Astrophysical Journal Supplement Series*, vol. 9, 201–319 [doi:10.1086/190103].

Fowler, William A. and Fred Hoyle. (c. 1964). *Nucleosynthesis in Massive Stars and Supernovae*. Chicago: University of Chicago Press.

Frankenberry, Nancy K., editor and commentator. (2008). *The Faith of Scientists in Their Own Words*. Princeton: Princeton University Press.

Flynn, Tom, editor. (2007). *The New Encyclopedia of Unbelief*, foreword by Richard Dawkins. Amherst: Prometheus.

Gale, George. (1993). "Multiple Universes" in *Cosmology: Historical, Literary, Philosophical, Religious, and Scientific Perspectives*, edited by N. S. Hetherington. New York: Garland; Routledge.

Gale, George and John R. Urani. (1993). "Philosophical Aspects of Cosmology" in *Cosmology: Historical, Literary, Philosophical, Religious, and Scientific Perspectives*, edited by N. S. Hetherington. New York: Garland; Routledge, 2008.

Gamow, George. (1946). "Expanding Universe and the Origin of Elements" in *Physical Review*, vol. 70, issue 7–8, 572–73 [doi:10.1103/PhysRev.70.572.2].

Gamow, George. (1948). "The Origin of Elements and the Separation of Galaxies" in *Physical Review*, vol. 74, issue 4, 505–6 [doi:10.1103/PhysRev.74.505.2].

Gamow, George. (October 30, 1948). "Evolution of the Universe" in *Nature*, vol. 162, 680–82 [doi:10.1038/162680a0].

Gamow, George. (July 1951). "The Origin and Evolution of the Universe" in *American Scientist*, vol. 39, issue 3, 393–407.

Gamow, George. (1961 [1952]). *The Creation of the Universe*. New York: Viking.

Gamow, George. (March 1954). "Modern Cosmology" in *Scientific American*, vol. 190, 54–63 [doi:10.1038/scientificamerican0354-54].

Gamow, George. (1954 [November 1953]). "On the Steady State Theory of the Universe" in *The Astronomical Journal*, vol. 59, issue 1217, 200 [doi:10.1086/106997].

Gamow, George. (September 1956). "The Evolutionary Universe" in *Scientific American*, vol. 192, 136–54 [doi:10.1038/scientificamerican0956-136].

Gamow, George. (1956). "The Physics of the Expanding Universe" in *Vistas in Astronomy*, vol. 2, issue 1, 1726–32 [doi:10.1016/0083-6656(56)90099-x].

Gamow, George. (1970). *My World Line: An Informal Autobiography*, foreword by Stanislaw M. Ulam. New York: Viking.

Gamwell, Franklin I. (1990). *The Divine Good: Modern Moral Theory and the Necessity of God.* San Francisco: Harper; SMU, 1996.

Gamwell, Franklin I. (2011). *Existence and the Good: Metaphysical Necessity in Morals and Politics.* Albany: State University of New York Press.

Gardner, Martin. (2003). *Are Universes Thicker than Blackberries: Discourses on Godel, Magic Hexagrams, Little Red Riding Hood, and Other Mathematical and Pseudoscientific Topics?* New York: Norton.

Gare, Arran. (2009). "Transcending the Disciplinary Boundaries" in *Cosmos and History: Journal of Nature and Social Philosophy,* vol. 5, no. 2, online at www.cosmosandhistory.org.

Gates, S. James Jr. (2006a). *Superstring Theory: The DNA of Reality, Part 1 of 2.* Chantilly: Teaching Company.

Gates, S. James Jr. (2006b). *Superstring Theory: The DNA of Reality, Part 2 of 2.* Chantilly: Teaching Company.

Gates, S. James Jr. (October 5, 2006). "Modern Cosmology and Superstring Theory: Can They Coexist?" Lecture at Concordia University, Montreal.

Gates, S. James Jr. (October 18, 2009). "Does Reality Have a Genetic Basis?" Quantum to Cosmos Festival Lecture at Perimeter Institute, Waterloo, Ontario [via YouTube, July 2013].

Gibson, Carl H. (2013). "Fluid Mechanics Explains Cosmology, Dark Matter, Dark Energy, and Life" in *Journal of Cosmology,* vol. 22, issue 3, 10012–24.

Gibson, Carl H. with Rudolph E. Schild and Nalin Chandra Wickramasinghe. (2010 [April 4]). "Origin of Life from Primordial Planets" in *International Journal of Astrobiology,* vol. 10, issue 2, 83–98 [doi:10.1017/S1473550410000352].

Gibson, Daniel G. with John I. Glass and others. (2010). "Creation of a Bacterial Cell Controlled by a Chemically Synthesised Genome" in *Science,* vol. 329, issue 5987, 52–56.

Gibson, Karl and Mariano Artigas. (2007). *Oracles of Science: Celebrity Scientists versus God and Religion.* Oxford: Oxford University Press.

Gilkey, Landon. (1993). *Nature, Reality, and the Sacred: The Nexus of Science and Religion.* Minneapolis: Fortress.

Gilroy, Paul. (1993). *The Black Atlantic: Modernity and Double Consciousness.* Cambridge: Harvard University Press.

Gingerich, Owen. (1984). *Astrophysics and Twentieth-Century Astronomy to 1950.* Cambridge: Cambridge University Press.

Gingerich, Owen. (1989). *The Physical Sciences in the Twentieth Century.* New York: Scribner.

Gingerich, Owen. (1992). "Response to Mortimer J. Adler" in *Great Ideas Today.* Chicago: Encyclopedia Britannica [response to Adler's "Natural Theology, Chance, and God" (Gingerich 2014: 158)].

Gingerich, Owen. (c. 1993). *The Eye of Heaven: Ptolemy, Copernicus, Kepler.* New York: American Institute of Physics.

Gingerich, Owen. (2004). *The Book Nobody Read: Chasing the Revolutions of Nicolaus Copernicus.* New York: Walker.

Gingerich, Owen. (2006). *God's Universe*, foreword by Peter J. Gomes. Cambridge: Harvard University Press.

Gingerich, Owen. (2014). "Was Hoyle Right?" in *God's Planet*, foreword by Randy Isaac. Cambridge: Harvard University Press.

Gingerich, Owen and James MacLachian. (2005). *Nicolaus Copernicus: Making the Earth a Planet*. Oxford: Oxford University Press.

Girardot, Norman and Mac Linscott Ricketts, editors. (1982). *Imagination and Meaning: The Scholarly and Literary Worlds of Mircea Eliade*. New York: Seabury.

Glendenning, Norman K. (2007). *Our Place in the Universe*. Hackensack: World Scientific.

Goheen, John. (1941). "Whitehead's Theory of Value" in *The Philosophy of Alfred North Whitehead, The Library of Living Philosophers, Vol. 3*, edited by P. A. Schilpp. La Salle: Open Court.

Gold, Thomas. (1953). "Relation between Modern Cosmologies and Nuclear Astrophysical Processes" in *Mémoires de la Societé Royale des Sciences de Liegè*, 4th series, vol. 13, 68–69.

Gold, Thomas. (February 26, 1955). "The 'Horizon' of the Steady-State Universe" in *Nature*, vol. 175, issue 4452, 382 [doi:10.1038/175382a0].

Gold, Thomas. (1999). *Deep Hot Biosphere*, foreword by Freeman Dyson. New York: Copernicus.

Gold, Thomas. (2005). *Getting the Back off the Watch and Other Episodes in the Life of a Scientist*. Oxford: Oxford University Press.

Gold, Thomas (assisted by D. L. Schumacher), editor. (1967). *The Nature of Time*. Ithaca: Cornell University Press.

Goldsmith, Donald. (February 14, 2012). "Far, Far Future of Stars" in *Scientific American*, vol. 306, 32–39 [doi:10.1038/scientificamerican0312–32].

Gonzalez-Mestres, Luis. (December 12, 2012). "Pre-Big Bang, Vacuum and Noncyclic Cosmoloiges" presented at the *2011 Europhysics Conference on High Energy Physics*, July 21–27, Grenoble, Rhone-Alpes, France [arXiv:1 212.2925v1].

Goodenough, Ursula. (1998). *The Sacred Depths of Nature*. New York: Oxford University Press.

Goodman, Steve. (2012). "The Ontology of Vibration Force" in *Sound Studies Reader*, edited by J. Sterne. New York: Routledge.

Gough, Douglas, editor. (2005). *The Scientific Legacy of Fred Hoyle*, foreword by Martin Rees. Cambridge: Cambridge University Press, 2011 paperback.

Gould, Stephen Jay. (October 1994). "The Evolution of Life on the Earth" in *Scientific American*, vol. 271, issue 4, 84–91.

Gragg, Alan. (1973). *Charles Hartshorne: Makers of the Modern Theological Mind*. Waco: Word.

Green, Michael. (2013). *Quantum Physics and Ultimate Reality: Mystical Writings of Great Physicists*. Amazon Kindle Direct Publishing.

Greene, Brian. (2004). *The Fabric of the Cosmos: Space, Time, and the Texture of Reality*. New York: Knopf.

Greene, John C. (1961). *Darwin and the Modern World View.* Baton Rouge: Louisiana State University Press.

Greene, John C. (1981). *Science, Ideology, and World View: Essays in the History of Evolutionary Ideas.* Berkeley: University of California Press.

Greene, V. W. with P. D. Pederson, D. A. Lundgren, C. A. Hagberg, and G. A. Soffen. (April 1964). "Microbiological Exploration of Stratosphere: Results of Six Experimental Flights" in *Proceedings of the Atmospheric Biology Conference* at University of Minnesota (April 13–15), 199–211. Washington: NASA.

Gregersen, Niels Henrik and J. Wentzel van Huyssteen. (1998). *Rethinking Theology and Science: Six Models for the Current Dialogue.* Grand Rapids: Eerdmans.

Gregersen, Niels Henrik with Ulf Görman and Christoph Wassermann, editors. (1999). *The Interplay between Scientific and Theological Worldviews.* Geneva: Labor et Fides.

Gregory, Jane. (January 2003). "The Popularization and Excommunication of Fred Hoyle's 'Life-from-Space' Theory" in *Public Understanding of Science,* vol. 12, issue 1, 25–46.

Gregory, Jane. (2005). *Fred Hoyle's Universe.* Oxford: Oxford University Press.

Gregory, Jane and Steve Miller. (1998). *Science in Public: Communication, Culture, and Credibility.* New York: Plenum.

Grenz, Stanley J. (1996). *A Primer on Postmodernism.* Grand Rapids: Eerdmans.

Grevesse-Guillaume, C. and N. C. Wickramasinghe. (1966 [July 21, 1965]). "On the Optics of Small Graphite Spheres, III" in *Monthly Notices of the Royal Astronomical Society,* vol. 132, issue 4, 471–73.

Gribbin, John and Martin Rees. (1989). *Cosmic Coincidences: Dark Matter, Mankind, and Anthropic Cosmology.* New York: Bantam.

Griffin, David Ray. (1989). *God and Religion in the Postmodern World: Essays in Postmodern Theology.* Albany: State University of New York Press.

Griffin, David Ray. (1992). "Creativity and Postmodern Religion" in *The Postmodern Reader,* edited by C. Jencks. New York: St. Martin's.

Griffin, David Ray. (1997). *Parapsychology, Philosophy and Spirituality: A Postmodern Exploration.* Albany: State University of New York Press.

Griffin, David Ray. (2000). *Reenchantment without Supernaturalism: A Process Philosophy of Religion.* Ithaca: Cornell University Press.

Griffin, David Ray. (2000). *Religion and Scientific Naturalism: Overcoming the Conflicts.* Albany: State University of New York Press.

Griffin, David Ray. (2003). "Reconstructive Postmodern Theology" in *The Cambridge Companion to Postmodern Theology,* edited by K. J. Vanhooer. Cambridge: Cambridge University Press.

Griffin, David Ray. (2004). "Panentheism: A Postmodern Revelation" in *In Whom We Live and Move and Have Our Being: Panentheistic Reflections on God's Presence in a Scientific World,* edited by P. Clayton and A. Peacocke. Grand Rapids: Eerdmans.

Griffin, David Ray. (2007). *Whitehead's Radically Different Postmodern Philosophy: An Argument for its Contemporary Relevance.* Albany: State University of New York Press.

Griffin, David Ray. (2008). "Whitehead's Naturalism and a Non-Darwinian View of Evolution" in *Back to Darwin: A Richer Account of Evolution*, edited by John Cobb. Grand Rapids: Eerdmans.

Griffin, David Ray. (2014). *Panentheism and Scientific Naturalism: Rethinking Evil, Morality, Religious Experience, Religious Pluralism, and the Academic Study of Religion* (Toward Ecological Civilization Series). Claremont: Process Century.

Griffin, David Ray, editor. (1988). *The Reenchantment of Science: Postmodern Proposals.* Albany: State University of New York Press.

Griffin, David Ray, editor. (1988). *Spirituality and Society: Postmodern Visions.* Albany: State University of New York Press.

Griffin, David Ray with William A. Beardslee and Joe Holland. (1989). *Varieties of Postmodern Theology.* Albany: State University of New York Press.

Griffith, Paul J. (2014). *Decreation: The Last Things of All Creatures.* Waco: Baylor University Press.

Grimm, R. E. and H. Y. McSween Jr. (1989). "Water and the Thermal Evolution of Carbonaceous Chondrite Parent Bodies" in *Icarus*, vol. 82, 244–80.

Grinnell, Frederick. (2009). *Everyday Practice of Science: Where Intuition and Passion Meet Objectivity and Logic.* Oxford: Oxford University Press.

de Groot, Natasjaa G. with Nel Otting and others. (2002). "Evidence for an Ancient Selective Sweep in the MHC Class I Gene Repertoire of Chimpanzees" in *Proceedings of the National Academy of Sciences*, vol. 99, issue 18, 11748–53.

Gunton, Colin E. (2001). *Becoming and Being: The Doctrine of God in Charles Hartshorne and Karl Barth.* Norwich: SCM.

Guth, Alan H. (January 15, 1981 [August 11, 1980]). "Inflationary Universe: A Possible Solution to the Horizon and Flatness Problems" in *Physical Review D*, vol. 23, issue 2, 347–56. [doi:10.1103/PhysRevD.23.347]

Guth, Alan H. (1993). "The Inflationary Universe" in *Cosmology: Historical, Literary, Philosophical, Religious, and Scientific Perspectives*, edited by N. S. Hetherington. New York: Garland; Routledge, 2008.

Guth, Alan H. (June 22, 2007). "Eternal Inflation and Its Implications" in *Journal of Physics A: Mathematical and Theoretical*, vol. 40, issue 25, 6811–26. [doi:10.1088/1751–8113/40/25/S25]

Guth, Alan H. and S.-Y. Pi. (October 11, 1982). "Fluctuations in the New Inflationary Universe" in *Physical Review Letters*, vol. 49, 1110–13 [doi:10.1103/PhysRevLett.49.1110].

Guth, Alan H. and Paul J. Steinhardt. (May 1984). "The Inflationary Universe" in *Scientific American*, vol. 250, issue 5, 116–28 [doi:10.1038/scientificamerican0584–116].

Gyatso, Tenzin (His Holiness the Fourteenth Dalai Lama). (2005). *The Universe in a Single Atom: The Convergence of Science and Spirituality.* New York: Doubleday.

Habermas, Jürgen. (1987). *The Philosophical Discourse of Modernity: Twelve Lectures*, translation by F. Lawrence. Cambridge: MIT.

Habermas, Jürgen. (1992). "Modernity: An Unfinished Project" in *Postmodern Reader*, edited by C. Jencks. New York: St. Martin's.

Haisch, Bernard. (2006). *The God Theory: Universes, Zero-Point Fields and What's Behind It All*. Newburyport: Weiser.

Haisch, Bernard. (2010). *The Purpose-Guided Universe: Believing in Einstein, Darwin, and God*. Franklin Lakes: New Page.

Haldane, J. B. S. (1929). *The Origin of Life*. London: Chatto and Windys.

Halvorson, Hans and Helge Kragh. (Fall 2013 [October 24, 2011]). "Cosmology and Theology" in *Stanford Encyclopedia of Philosophy*, edited by E. N. Zalta, online at http://plato.stanford.edu/archives/fall2013/entries/cosmology-theology/.

Hameroff, Stuart and Roger Penrose. (March 2014). "Consciousness in the Universe: A Review of the 'Orch OR' Theory" in *Physics of Life Reviews*, vol. 11, issue 1, 39–78 [doi:10.1016/j.plrev.2013.08.002].

Hancock, Roger. (1972 [1967]). "Metaphysics, History of It" in *The Encyclopedia of Philosophy, Vol. 5*, edited by P. Edwards. New York: Macmillan.

Hand, Eric. (January 12, 2012). "Radio Array Starts Work: Giant Low-Frequency Sensor System on Track to Probe the Birth of the First Stars" in *Nature News*, vol. 481, issue 738, 127 [doi:10.1038/481127a].

Hanna, Judith Lynne. (1987). *To Dance is Human: A Theory of Nonverbal Communications*. Chicago: University of Chicago Press.

Hannay, J. B. (February 15, 1883). "Natural Selection and Natural Theology" in *Nature*, vol. 27, 364 [doi:10.1038/027364a0].

Harley, Clifford. (1935). *God and the Universe: A Sympathetic Examination of the Book by Sir James Jeans's "The Mysterious Universe."* Manchester: New-Church Publishing Society.

Harris, Mark. (2014). *The Nature of Creation: Examining the Bible and Science*. New York: Routledge.

Harrison, Edward. (2005 [1981]). *Cosmology: The Science of the Universe*. Cambridge: Cambridge University Press.

Harrison, Peter, editor. (2010). *Cambridge Companion to Science and Religion*. Cambridge: Cambridge University Press.

Hart, Trevor. (2014). *Making Good: Creation, Creativity, and Artistry*. Waco: Baylor University Press.

Hartman, Robert S. (2011 [1965]). *The Structure of Value: Foundations of Scientific Axiology*, foreword by G. Kimball Plochmann, preface by P. Weiss. Eugene: Wipf and Stock.

Hartogh, Paul with Dariusz C. Lis, Dominique Bockelée-Morvan, and others. (October 13, 2011 [May 16]). "Ocean-like Water in the Jupiter-family Comet 103P/Hartley 2" in *Nature*, vol. 478, issue 7368, 218–20 [doi:10.1038/nature10519].

Hartshorne, Charles. (1923). *An Outline and Defense of the Argument for the Unity of Being in the Absolute or Divine Good*. Harvard PhD Dissertation.

Hartshorne, Charles. (1934). *The Philosophy and Psychology of Sensation.* Chicago: University of Chicago Press; Port Washington: Kennikat, 1968.

Hartshorne, Charles. (1936). "The Compound Individual" in *Philosophical Essays for Alfred North Whitehead*, edited by Otis H. Lee. London: Longman's Green.

Hartshorne, Charles. (1937). "The Cosmic Variables" is chapter VIII (111–24) in *Beyond Humanism: Essays in the Philosophy of Nature.* Chicago: Willett, Clark & Company; with important new preface, Gloucester: Peter Smith, 1975.

Hartshorne, Charles. (1941a). "God and Righteousness" is chapter IV (142–73) in *Man's Vision of God and the Logic of Theism.* Chicago: Willett; Hamden: Archon, 1964.

Hartshorne, Charles. (1941b). "Whitehead's Idea of God" in *The Philosophy of Alfred North Whitehead, The Library of Living Philosophers, Vol. 3*, edited by P. A. Schilpp. La Salle: Open Court.

Hartshorne, Charles. (May 1944). "The Formal Validity and the Real Significance of the Ontological Argument" in *Philosophical Review*, vol. 53, 225–45.

Hartshorne, Charles (1948). *The Divine Relativity: A Social Conception of God.* New Haven: Yale University Press.

Hartshorne, Charles. (1950). "Whitehead's Metaphysics" in *Whitehead and the Modern World: Science, Metaphysics, and Civilization—Three Essays on the Thought of Alfred North Whitehead* by Victor Lowe, Charles Hartshorne, and A. H. Johnson, preface by A. C. Benjamin. Boston: Beacon.

Hartshorne, Charles. (1953). *Reality as Social Process: Studies in Metaphysics and Religion*, foreword by William Ernest Hocking. Boston: Beacon; New York: Hafner, 1971.

Hartshorne, Charles. (August 1961). "The Logic of the Ontological Argument" in *Journal of Philosophy*, vol. 58, 471–73.

Hartshorne, Charles. (1962). *The Logic of Perfection and Other Essays in Neoclassical Metaphysics.* La Salle: Open Court.

Hartshorne, Charles. (1965). *Anselm's Discovery: A Re-Examination of the Ontological Proof for God's Existence.* La Salle: Open Court.

Hartshorne, Charles. (1967). *A Natural Theology for Our Time.* La Salle: Open Court.

Hartshorne, Charles. (1970). "A Philosophy of Shared Creative Experience" and "What Metaphysics Is" in *Creative Synthesis and Philosophic Method.* La Salle: Open Court.

Hartshorne, Charles. (1971). "The Development of Process Philosophy" in *Process Theology: Basic Writings by the Key Thinkers of a Major Modern Movement*, edited by E. Cousins. New York: Newman.

Hartshorne, Charles. (1972). *Whitehead's Philosophy; Selected Essays, 1935–1970.* Lincoln: University of Nebraska Press.

Hartshorne, Charles. (1973). *Born to Sing: An Interpretation and World Survey of Bird Song.* Bloomington: University of Indiana Press.

Hartshorne, Charles. (1976). *Aquinas to Whitehead: Seven Centuries of Metaphysics of Religion* (Aquinas Lecture). Milwaukee: Marquette University.

Hartshorne, Charles. (Spring 1976). "Why Psychicalism?" in *Process Studies*, vol. 6, issue 1, 67–72.

Hartshorne, Charles. (Fall 1977). "Bell's Theorem and Stapp's Revised View of Space-Time" in *Process Studies*, vol. 7, 183–91.

Hartshorne, Charles. (1981). "The Ethics of Contributionism" in *Responsibilities to Future Generations: Environmental Ethics*, edited by E. Partridge. Buffalo: Prometheus.

Hartshorne, Charles. (1983). *Insights and Oversights of Great Thinkers: An Evaluation of Western Philosophy*. Albany: State University of New York Press.

Hartshorne, Charles (1984a). *Creativity in American Philosophy*. Albany: State University of New York Press.

Hartshorne, Charles. (1984b). "The Physical and the Spiritual" in *Omnipotence and Other Theological Mistakes*. Albany: State University of New York Press.

Hartshorne, Charles. (1987a). "A Metaphysics of Universal Freedom" in *Faith and Creativity: Essays in Honor of Eugene H. Peters*, edited by G. Nordgulen and G. W. Shields. St. Louis: CBP Press.

Hartshorne, Charles (1987b). "Two Views of Metaphysics" in *Wisdom as Moderation: A Philosophy of the Middle Way*, edited by Robert C. Neville. Albany: State University of New York Press.

Hartshorne, Charles. (1990). *The Darkness and the Light: A Philosopher Reflects upon His Fortunate Career and Those Who Made it Possible*. Albany: State University of New York Press.

Hartshorne, Charles. (1991). "An Open Letter to Carl Sagan" in *Journal of Speculative Philosophy*, vol. 5, 227–32.

Hartshorne, Charles. (1997). *The Zero Fallacy and Other Essays in Neoclassical Philosophy*, edited by Mohammad Valady. La Salle: Open Court.

Hartshorne, Charles. (May 2001). "Redefining God" in *American Journal of Theology and Philosophy*, vol. 22, issue 2, 107–13.

Hartshorne, Charles. (2011). *Creative Experiencing: A Philosophy of Freedom*, edited by Donald Wayne Viney and O. Jincheol. Albany: State University of New York Press.

Hartshorne, Charles and William L. Reese. (1953). *Philosophers Speak of God*. Chicago: University of Chicago Press; Midway, 1976; Humanity, 2000.

Hartshorne, Charles and Creighton Peden. (1981). *Whitehead's View of Reality*. New York: Pilgrim.

Haselgrove, C. B. and F. Hoyle. (May 10, 1956). "A Preliminary Determination of the Age of Type II Stars" in *Monthly Notices of the Royal Astronomical Society*, vol. 116, issue 5, 527–32.

Hattche, Matt. (Fall 2008). "Christian Wolff" in *Stanford Encyclopedia of Philosophy*. Stanford: Stanford University.

Hauerwas, Stanley. (2000). "On Keeping Theological Ethics Theological" in *The Hauerwas Reader*, edited by John Berkman and Michael G. Cartwright. Durham: Duke University Press.

Hauerwas, Stanley. (2001). *With the Grain of the Universe: The Church's Witness and Natural Theology: Being the Gifford Lectures Delivered at the University of St. Andrews in 2001.* Grand Rapids: Brazos.

Hauerwas, Stanley with Nancey Murphy and Mark Nation, editors. (1994). *Theology without Foundations: Religious Practice and the Future of Theological Truth.* Nashville: Abingdon.

Haught, John F. (2000). *God after Darwin: A Theology of Evolution.* Boulder: Westview.

Haught, John F. (2003). *Deeper than Darwin: The Prospect for Religion in the Age of Evolution.* Boulder: Westview.

Haught, John F. (2008). *God and the New Atheism: A Critical Response to Dawkins, Harris, and Hitchens.* Louisville: Westminster John Knox.

Hawking, Stephen W. (1988). *A Brief History of Time: From the Big Bang to Black Holes.* New York: Bantam.

Hawking, Stephen W. and Leonard Mlodinow. (2005). *A Briefer History of Time.* New York: Bantam.

Hawking, Stephen W. and Leonard Mlodinow. (2010). *Grand Design.* New York: Bantam.

Hawking, Stephen W. and Roger Penrose. (2010 [1996]). *The Nature of Space and Time*, with new afterword. Princeton: Princeton University Press.

Hawking, Stephen W., editor and commentator. (2002). *On the Shoulders of Giants: The Great Works of Physics and Astronomy.* Philadelphia: Running Press.

Hawking, Stephen W., editor and commentator. (2005). *God Created the Integers: The Mathematical Breakthroughs that Changed History.* Philadelphia: Running Press.

Heavens, Alan. (25 November 2010). "Cosmology: Geometry of the Universe" in *Nature*, vol. 468, issue 7323, 511–12 [doi:10.1038/468511a].

Hedges, Chris. (2008). *I Don't Believe in Atheists: When Atheism Becomes Religion: America's New Fundamentalists.* New York: Free Press.

Heller, Michael. (2003). *Creative Tension: Essays on Science and Religion.* Philadelphia: Templeton.

Henning, Brian G. (2005). *The Ethics of Creativity: Beauty, Morality, and Nature in a Processive Cosmos.* Pittsburg: University of Pittsburg Press.

Herrmann, Robert L., editor. (2000). *God, Science, and Humility: Ten Scientists Consider Humility Theology.* Philadelphia: Templeton.

Herrmann, Robert L., editor (2001). *Expanding Humanity's Vision of God: New Thoughts on Science and Religion.* Philadelphia: Templeton.

Herzfeld, Noreen L., (2009). *Technology and Religion: Remaining Human in a Co-created World.* West Conshohocken: Templeton.

Hesburgh, Theodore M. (1956 [1950]). "Attributes of God" in *God and the World of Man.* Notre Dame: University of Notre Dame Press.

Hetherington, Edith W. and Norriss S. Hetherington. (2009). *Astronomy and Culture.* Santa Barbara: Greenwood.

Hetherington, Norriss. (1987). *Ancient Astronomy and Civilization*, illustrations by V. A. Mann. Tucson: Pachart.

Hetherington, Norriss S. (1988). *Science and Objectivity: Episodes in the History of Astronomy*. Ames: Iowa State University Press.

Hetherington, Norriss S., editor. (1993). *Cosmology: Historical, Literary, Philosophical, Religious, and Scientific Perspectives*. New York: Garland; Routledge, 2008.

Hetherington, Norriss S., editor. (1993). *Encyclopedia of Cosmology: Historical, Philosophical, and Scientific Foundations of Modern Cosmology*. New York: Garland; Routledge, 2008.

Hewitt, Adelaide with Geoffrey Burbidge and Li Zhi Fang. (1987). *Observational Cosmology: Proceedings of the 124th Symposium of the International Astronomical Union*, August 25–30, 1986, Beijing, China. Dordrecht: D. Reidel.

Hey, J. S. (1971). *The Radio Universe*. New York: Pergamon; second edition 1976; third edition 1983.

Hey, J. S. (1973). *The Evolution of Radio Astronomy*. New York: Science History.

Hey, J. S. with S. J. Parsons and J. W. Phillips. (August 17, 1946). "Fluctuations in Cosmic Radiation at Radio-Frequencies" in *Nature*, vol. 158, 234 [doi:10.1038/158234a0].

Higginson, Edward. (1855). *Astro-theology*. London: E. T. Whitfield.

Hoag, Hannah (December 17, 2014). "Earth's Deep Crust Could Support Widespread Life: Discovery of Hydrogen-rich Waters Hints at Unexploded Microbial Ecosystems" in *Nature News* [doi:10.1038/nature.2014.16575].

Hodgkin, Alan Lloyd. (1964). *The Conduction of the Nervous Impulse*. Springfield: C. C. Thomas.

Hogue, W. Lawrence. (2013). *Postmodernism, Traditional Cultural Forms, and African American Narratives*. Albany: State University of New York Press.

Holder, Rodney D. (2004). *God, the Multiverse, and Everything: Modern Cosmology and the Argument from Design*. Aldershot: Ashgate.

Holder, Rodney D. (2012). *The Heavens Declare: Natural Theology and the Legacy of Karl Barth*. West Conshohocken: Templeton.

Holder, Rodney D. (2013). *Big Bang Big God: A Universe Designed for Life*. Oxford: Lion Hudson.

Holder, Rodney D. and Simon Mitton, editors. (2012). *Georges Lemaître: Life, Science, and Legacy*. Berlin: Springer-Verlag.

Holmes, Barbara A. (2002). *Race and the Cosmos: An Invitation to View the World Differently*. Harrisburg: Trinity.

Holmes, Barbara A. (2008). *Liberation and the Cosmos: Conversations with the Elders*. Minneapolis: Fortress.

Holton, Gerald and Stephen G. Brush. (2001). *Physics, the Human Adventure: From Copernicus to Einstein and Beyond*, third edition. New Brunswick: Rutgers University Press.

Hoover, Richard B. (February–March 2011). "Fossils of Cyanobacteria in CI1 Carbonaceous Meteorites" in *Journal of Cosmology*, vol. 13, issue 35, 3811–48 [And "Implications for Life on Comets, Europa, and Enceladus" vol. 15, II.e, 6249–87 (July–August 2011), online at journalofcosmology.com/Contents15_files/Hoover_JOC_MS.pdf.

Hoover, Richard B. with Fred Hoyle, N. C. Wickramasinghe, Miriam J. Hoover, and S. Al-Mufti. (April 1986). "Diatoms on Earth, Comets, Europa, and in Interstellar Space" in *Earth, Moon and Planets*, vol. 35, issue 1, 19–45.

Hoover, Richard B. with A. Yu Rozanov and Roland Paepe, editors. (2005). *Perspectives in Astrobiology*, Amsterdam: IOS Press.

Horie, Masayuki with Tomoyuki Honda, Yoshiyuki Suzuki, and others. (January 7, 2010). "Endogenous Non-retroviral RNA Virus Elements in Mammalian Genomes" in *Nature*, vol. 463, issue 7277, 84–87.

Hosinski, Thomas E. (1993). *Stubborn Fact and Creative Advance: An Introduction to the Metaphysics of Alfred North Whitehead*. Lanham: Rowman and Littlefield.

Hoskin, Michael, editor. (1997). *The Cambridge Illustrated History of Astronomy*. Cambridge: Cambridge University Press.

Hoskin, Michael, editor. (1999). *The Cambridge Concise History of Astronomy*. Cambridge: Cambridge University Press.

Hoskin, Michael. (2003). *History of Astronomy: A Very Short Introduction*. Oxford: Oxford University Press.

Houts, Michael G. (November 2007). "Evolution Is a Religion—Not Science [Part I]." *Reason and Revelation: A Monthly Journal of Christian Evidences*, vol. 27, issue 11, 81–87.

Howell, Alfred H. (1989). *Who Made You?: Theology, Science, and Human Responsibility*. New York: Praeger.

Howell, Nancy R. (2000). *A Feminist Cosmology: Ecology, Solidarity, and Metaphysics*. Amherst: Humanity.

Hoyle, Fred. (1939). "Quantum Electrodynamics" in *Mathematical Proceedings of the Cambridge Philosophical Society*, vol. 35, issue 3, 419–62.

Hoyle, Fred. (December 2, 1946a [April 6]). "The Synthesis of the Elements from Hydrogen" in *Monthly Notices of the Royal Astronomical Society*, vol. 106, issue 5, 343–83.

Hoyle, Fred. (April 12, 1946b). "On the Condensation of the Planets" in *Monthly Notices of the Royal Astronomical Society*, vol. 106, issue 5, 406–22.

Hoyle, Fred. (December 29, 1946c). "The Chemical Composition of the Stars" in *Monthly Notices of the Royal Astronomical Society*, vol. 106, issue 4, 255–59.

Hoyle, Fred. (1947). "On the Formation of Heavy Elements in Stars" in *Proceedings of the Physical Society*, vol. 59, issue 6, 972–78.

Hoyle, Fred. (August 5, 1948). "A New Model for the Expanding Universe" in *Monthly Notices of the Royal Astronomical Society*, vol. 108, issue 1748, 372–82.

Hoyle, Fred. (1949). *Some Recent Researches in Solar Physics*. Cambridge: Cambridge University Press.

Hoyle, Fred. (January 14, 1949a). "On the Cosmological Problem" in *Monthly Notices of the Royal Astronomical Society*, vol. 109, issue 3, 365–71.

Hoyle, Fred. (February 5, 1949b). "Stellar Evolution and the Expanding Universe" in *Nature*, vol. 163, 196–98 [doi:10.1038/163196a0].

Hoyle, Fred. (1950). *The Nature of the Universe: A Series of Broadcast Lectures.* Oxford: B. Blackwell.

Hoyle, Fred. (June 1, 1952). "Concepts of the Universe" in *New York Times Magazine*, 11–12, 50–51. (Kragh 1996: 463)

Hoyle, Fred. (1953a). *A Decade of Decision*. London: Heinemann.

Hoyle, Fred. (August 15, 1953b). "Cosmic Origin of Radiation at Radio Frequencies" in *Nature*, vol. 172, issue 4372, 296–97 [doi:10.1038/172296a0].

Hoyle, Fred. (November 1953). "On the Fragmentation of Gas Clouds Into Galaxies and Stars" in *The Astrophysical Journal*, vol. 118, 513–28 [doi:10.1086/145780].

Hoyle, Fred. (March 13, 1954a). "Generation of Radio Noise by Cosmic Sources" in *Nature*, vol. 173, issue 4402, 483–84 [doi:10.1038/173483a0].

Hoyle, Fred. (September 1954b [December 22, 1953]). "On Nuclear Reactions Occurring in Very Hot Stars. I. The Synthesis of Elements from Carbon to Nickel" in *The Astrophysical Journal Supplement*, vol. 1, 121–46 [doi:10.1086/190005].

Hoyle, Fred. (1955a). *Frontiers of Astronomy.* New York: Harper; London: Heinemann [reprinted in 1956, 1961, and 1963; Spanish version in 1960 and 1970].

Hoyle, Fred. (May 7, 1955b). "The 'Horizon' of the Steady-State Universe" [reply to Thomas Gold, February 26, 1955] in *Nature*, vol. 175, issue 4462, 808 [doi:10.1038/175808a0].

Hoyle, Fred. (September 1956a). "The Steady-State Universe" in *Scientific American*, vol. 195, 157–67 [doi:10.1038/scientificamerican0956–157].

Hoyle, Fred. (1956b). *Man and Materialism.* New York: Harper.

Hoyle, Fred. (1957). *The Black Cloud.* London: Heinemann.

Hoyle, Fred. (1959). *Rockets in Ursa Major: A Novel.* London: Heinemann.

Hoyle, Fred. (1960a). *The Nature of the Universe.* New York: Harper.

Hoyle, Fred. (1960b). "On the Origin of the Solar Nebula" in *Quarterly Journal of the Royal Astronomical Society*, vol. 1, issue 1, 28–55.

Hoyle, Fred. (1961). "Observational Tests in Cosmology" [Guthrie Lecture] in *Proceedings of the Physical Society*, vol. 77, issue 1, 1–16 [doi:10.1088/0370–1328/77/1/302].

Hoyle, Fred. (1962). *Astronomy.* Garden City: Doubleday.

Hoyle, Fred. (1963). "Formation of the Planets" in *Origin of the Solar System: Proceedings of a Conference Held at the Goddard Institute for Space Studies*, January 23–24, 1962, New York, edited by Robert Jastrow and A. G. W. Cameron. New York: Academic.

Hoyle, Fred. (1965a). *Galaxies, Nuclei and Quasars.* New York: Harper.

Hoyle, Fred. (October 9, 1965b). "Recent Developments in Cosmology" in *Nature*, vol. 208, issue 5006, 111–14 [doi:10.1038/208111a0].

Hoyle, Fred. (1968). "Review of Recent Developments in Cosmology" [Bakerian Lecture] in *Proceedings of the Royal Society of London*, vol. 308, issue 1492, 1–17 [doi:10.1098/rspa.1968.0204].

Hoyle, Fred. (1972). *From Stonehenge to Modern Cosmology.* San Francisco: Freeman.

Hoyle, Fred. (1973a). *Nicolaus Copernicus: An Essay on His Life and Work*. London: Heinemann.
Hoyle, Fred. (1973b). "The Origin of the Universe" in *Quarterly Journal of the Royal Astronomical Society*, vol. 14, 278–87.
Hoyle, Fred. (1975a). *Astronomy and Cosmology: A Modern Course*. San Francisco: Freeman.
Hoyle, Fred. (1975b). *Astronomy Today*. London: Heinemann.
Hoyle, Fred. (1975c). *Highlights in Astronomy*. San Francisco: Freeman.
Hoyle, Fred. (March 15, 1975 [August 15, 1974]). "On the Origin of the Microwave Background" in *The Astrophysical Journal*, vol. 196, 661–70 [doi:10.1086/153452].
Hoyle, Fred. (1977a). *On Stonehenge*. London: Heinemann.
Hoyle, Fred. (1977b). *Ten Faces of the Universe*. London: Heinemann.
Hoyle, Fred. (1978). *Cosmogony of the Solar System*. Cardiff: University College Cardiff Press.
Hoyle, Fred. (1980a). *Steady-State Cosmology Revisited*. Cardiff: University College Cardiff Press.
Hoyle, Fred. (April 15, 1980b). *The Relation of Biology to Astronomy*. Cardiff: University College Cardiff Press.
Hoyle, Fred. (May 1981a). *The Universe: Past and Present Reflections*. Cardiff: University College Cardiff Press.
Hoyle, Fred. (November 1981b). "The Universe: Past and Present Reflections" in *Engineering and Science*, 8–12.
Hoyle, Fred. (November 12, 1981c). "Hoyle on Evolution" [Hoyle's famous "chance that a tornado sweeping through a junk-yard might assemble a Boeing 747" analogy for evolution by chance assembly] in *Nature*, vol. 294, issue 5837, 104–5 [doi:10.1038/294104b0].
Hoyle, Fred. (November 19, 1981d). "The Big Bang in Astronomy" in *New Scientist*, vol. 92, issue 1280, 521–27.
Hoyle, Fred. (1982a). "From Virus to Cosmology" audio online at www.buckingham.ac.uk/research/bcab/hrwarchieve.
Hoyle, Fred. (1982b). "The Universe: Past and Present Reflections" in *Annual Review of Astronomy and Astrophysics*, vol. 20, 1–35.
Hoyle, Fred. (1984 [c. 1983]). *The Intelligent Universe: A New View of Creation and Evolution*. New York: Holt, Rinehart and Winston.
Hoyle, Fred. (1985). *Comet Halley: A Novel in Two Parts*. New York: St. Martin's.
Hoyle, Fred. (1986). *The Small World of Fred Hoyle: An Autobiography*. London: M. Joseph.
Hoyle, Fred. (1989). "The Steady-State Theory Revived" in *Comments on Astrophysics*, vol. 13, issue 2, 81–86.
Hoyle, Fred. (May 4, 1989). "Articles of Faith" [review of *Science and Providence* (1989) by J. C. Polkinghorne] in *Nature*, vol. 339, issue 6219, 23 [doi:10.1038/339023a0].
Hoyle, Fred. (October 5, 1989). "What's in a Name?" in *Nature*, vol. 341, issue 6241, 380 [doi:10.1038/341380c0].

Hoyle, Fred. (April 26, 1990). "High Hopes for the Space Telescope" in *Nature*, vol. 344, commentary, 808–10 [doi:10.1038/344808a0].

Hoyle, Fred. (November 22, 1990). "Birth of the Gods" in *Nature*, vol. 348, issue 6299, 353–54 [doi:10.1038/348353a0].

Hoyle, Fred. (1993). "Anthropic and Perfect Cosmological Principles: Similarities and Differences" in *Anthropic Principle: Proceedings of the Second Venice Conference on Cosmology and Philosophy*, edited by F. Bertola and U. Curi. Cambridge: Cambridge University Press.

Hoyle, Fred. (1993). *The Origin of the Universe and the Origin of Religion*. Wakefield: Moyer Bell.

Hoyle, Fred. (April 15, 1993). "Heavenly Works" [review of *Nicholas Copernicus: Complete Works in Two Volumes* (1993) and *Johannes Kepler: New Astronomy* (1992)] in *Nature*, vol. 362, issue 6421, 657–58 [doi:10.1038/362657a0].

Hoyle, Fred. (1994). *Home Is Where the Wind Blows: Chapters from a Cosmologist's Life*. Mill Valley: University Science.

Hoyle, Fred. (1999). *Mathematics of Evolution*. Memphis: Acorn Enterprises.

Hoyle, Fred and N. C. Wickramasinghe. (June 13, 1962). "On Graphite Particles as Interstellar Grains" in *Monthly Notices of the Royal Astronomical Society*, vol. 124, 417–33.

Hoyle, Fred and N. C. Wickramasinghe. (July 8, 1963). "On the Deficiency in the Ultrviolet Fluxes from Early Type Stars" in *Monthly Notices of the Royal Astronomical Society*, vol. 126, issue 4, 401–4.

Hoyle, Fred and N. Chandra Wickramasinghe. (June 3, 1967). "Impurities in Interstellar Grains" in *Nature*, vol. 214, issue 5092, 969–71 [doi:10.1038/21496a0].

Hoyle, Fred and N. C. Wickramasinghe. (February 3, 1968). "Condensation of the Planets" in *Nature*, vol. 217, issue 5127, 415–18 [doi:10.1038/217415a0].

Hoyle, Fred and N. Chandra Wickramasinghe. (June 22, 1968). "Condensation of Dust in Galactic Explosions" in *Nature*, vol. 218, issue 5147, 1126–27 [doi:10.1038/2181126a0].

Hoyle, Fred and N. Chandra Wickramasinghe. (August 2, 1969a). "Interstellar Graphite and Silicates" in *Nature*, vol. 223, issue 5205, 445 [doi:10.1038/223445a0].

Hoyle, Fred and N. Chandra Wickramasinghe. (August 2, 1969b). "Interstellar Grains" in *Nature*, vol. 223, issue 5205, 459–62 [doi:10.1038/223459a0].

Hoyle, Fred and N. Chandra Wickramasinghe. (April 4, 1970). "Dust in Supernova Explosions" in *Nature*, vol. 226, issue 5240, 62–63 [doi:10.1038/226062a0].

Hoyle, Fred and N. C. Wickramasinghe. (August 1 [June 15], 1970). "Radio Waves from Grains in HII Regions" in *Nature*, vol. 227, issue 5257, 473–74 [doi:10.1038/227473a0].

Hoyle, Fred and N. Chandra Wickramasinghe. (November 4, 1976). "Primitive Grain Clumps and Organic Compounds in Carbonaceous Chondrites" in *Nature*, vol. 264, issue 5581, 45–46 [doi:10.1038/264045a0].

Hoyle, Fred and N. C. Wickramasinghe. (November [August 4] 1977). "Polysaccharides and the Infrared Spectrum of OH 26.5 + 0.6" in *Monthly Notices of the Royal Astronomical Society*, vol. 181, 51P–55P.

Hoyle, Fred and N. Chandra Wickramasinghe. (August 18, 1977). "Polysaccharides and the Infrared Spectra of Galactic Sources" in *Nature*, vol. 268, issue 5621, 610–12.

Hoyle, Fred and N. C. Wickramasinghe (November 24, 1977). "Identification of the λ2, 200Å Interstellar Absorption Feature" in *Nature*, vol. 270, issue 5635, 323–24 [doi:10.1038/270323a0].

Hoyle, Fred and N. C. Wickramasinghe. (December 22, 1977). "Origin and Nature of Carbonaceous Material in the Galaxy" in *Nature*, vol. 270, issue 5639, 701–3 [doi:10.1038/270701a0].

Hoyle, Fred and N. C. Wickramasinghe. (February 1978a [September 20, 1977]). "Calculations of Infrared Fluxes from Galactic Sources for a Polysaccharide Grain Model" in *Astrophysics and Space Science*, vol. 53, issue 2, 489–505.

Hoyle, Fred and Chandra Wickramasinghe. (February 1978b [December 19, 1977]). "Comets, Ice Ages and Ecological Catastrophes" in *Astrophysics and Space Science*, vol. 53, issue 2, 523–26.

Hoyle, Fred and Chandra Wickramasinghe. (1978c). *Lifecloud: The Origin of Life in the Universe*. London: J. M. Dent.

Hoyle, Fred and Chandra Wickramasinghe. (1979a). *Diseases from Space*. London: J. M. Dent.

Hoyle, Fred and Chandra Wickramasinghe. (1979b [March 30, 1979]). "On the Nature of Interstellar Grains" in *Astrophysics and Space Science*, vol. 66, issue 1, 77–90.

Hoyle, Fred and N. C. Wickramasinghe. (May 1980 [November 6, 1979]). "Organic Grains in Space" in *Astrophysics and Space Science*, vol. 69, issue 2, 511–13.

Hoyle, Fred and Chandra Wickramasinghe. (1981a [1980]). "Comets—a Vehicle for Panspermia" in *Comets and the Origin of Life: Proceedings of the Fifth College Park Colloquium on Chemical Evolution*, October 29–31, University of Maryland, edited by Cyril Ponnamperuma. Boston: Kluwer.

Hoyle, Fred and N. Chandra Wickramasinghe. (1981b). "Convergence to God" is chapter 9 in *Evolution from Space: A Theory of Cosmic Creationism*. New York: Simon and Schuster, 129–45.

Hoyle, Fred and N. Chandra Wickramasinghe. (1981c). *Space Travellers: The Bringers of Life*. Cardiff: University College Cardiff Press.

Hoyle, Fred and N. Chandra Wickramasinghe. (1982a). "Comets" is section C in *Proofs that Life Is Cosmic*. Memoirs of the Institute of Fundamental Studies. Sri Lanka: Government Press, 23–35.

Hoyle, Fred and N. Chandra Wickramasinghe. (1982b). *Is Life an Astronomical Phenomenon?* Cardiff: University College Cardiff Press.

Hoyle, Fred and N. Chandra Wickramasinghe. (1982c). *Why Neo Darwinism Does Not Work*. Cardiff: University College Cardiff Press.

Hoyle, Fred and N. Chandra Wickramasinghe. (September 8, 1983a). "Organic Grains in Taurus Interstellar Clouds" in *Nature*, vol. 305, issue 5930, 161 [doi:10.1038/305261a0].

Hoyle, Fred and N. Chandra Wickramasinghe. (December 1, 1983b). "Bacterial Life in Space" in *Nature*, vol. 306, issue 5942, 420 [doi:10.1038/306420a0].

Hoyle, Fred and N. Chandra Wickramasinghe. (1984). *From Grains to Bacteria*. Cardiff: University College Cardiff Press.

Hoyle, Fred and Chandra Wickramasinghe. (1985). *Living Comets*. Cardiff: University College Cardiff Press.

Hoyle, Fred and N. Chandra Wickramasinghe. (1986a). *Viruses from Space*. Cardiff: University College Cardiff Press.

Hoyle, Fred and N. C. Wickramasinghe. (August 7, 1986b). "The Case for Life as a Cosmic Phenomenon" in *Nature*, vol. 322, issue 6079, 509–11 [doi:10.1038/322509a0].

Hoyle, Fred and N. Chandra Wickramasinghe. (July 9, 1987). "Organic Dust in Comet Halley" in *Nature*, vol. 328, issue 6126, 117 [doi:10.1038/328117a0].

Hoyle, Fred and N. Chandra Wickramasinghe. (January 1988a). "The Organic Nature of Cometary Grains" in *Earth, Moon, and Planets*, vol. 40, issue 1, 101–8.

Hoyle, Fred and Chandra Wickramasinghe. (1988b). "The Concept of a Creator" is chapter 10 in *Cosmic Life-Force*. London: J. M. Dent; New York: Paragon, 1990, 132–44.

Hoyle, Fred and N. Chandra Wickramasinghe. (January 14, 1988c). "Cometary Organics" in *Nature*, vol. 331, issue 6152, 123–24 [doi:10.1038/331123c0].

Hoyle, Fred and N. Chandra Wickramasinghe. (February 25, 1988d). "Cometary Organics" in *Nature*, vol. 331, issue 6158, 666 [doi:10.1038/331/331666c0].

Hoyle, Fred with N. Chandra Wickramasinghe. (September 1989). "The Microwave Background in Steady-State Cosmology" in *Proceedings of the 22nd Eslab Symposium on Infrared Spectroscopy in Astronomy*, Salamanca, Spain, December 7–9, 1988, 489–95.

Hoyle, Fred and Chandra Wickramasinghe. (1990). *Cosmic Life-Force*. New York: Paragon.

Hoyle, Fred and N. C. Wickramasinghe. (April 1990). "Influenza—Evidence against Contagion" in *Journal of the Royal Society of Medicine*, vol. 83, issue 4, 258–61.

Hoyle, Fred and N. Chandra Wickramasinghe. (1991). *The Theory of Cosmic Grains*. Dordrecht: Kluwer.

Hoyle, Fred and N. Chandra Wickramasinghe. (1993). *Our Place in the Cosmos: The Unfinished Revolution*. London: J. M. Dent.

Hoyle, Fred and Chandra Wickramasinghe. (1997). *Life on Mars?: The Case for a Cosmic Heritage*, foreword by series editor Paul R. Goddard. Redland: Clinical.

Hoyle, Fred and N. C. Wickramasinghe. (October 1999). "Universe and Life: Deductions from the Weak Anthropic Principle" in *Astrophysics and Space Science*, vol. 268, issue 1–3, 89–102.

Hoyle, Fred and N. C. Wickramasinghe, editors. (2000). *Astronomical Origins of Life: Steps towards Panspermia*. Dordrecht: Kluwer.

Hoyle, Fred with N. Chandra Wickramasinghe and S. Al-Mufti. (1985b [October 29, 1984]). "Ultraviolet Absorbance of Presumably Interstellar Bacteria and Related Matters" in *Astrophysics and Space Science*, vol. 111, issue 1, 65–78.

Hoyle, Fred with N.C. Wickramasinghe and A. H. Olavesen. (January 19, 1978). "Identification of Interstellar Polysaccharides and Related Hydrocarbons" in *Nature*, vol. 271, issue 5642, 229–31.

Hoyle, Fred with N. C. Wickramasinghe and Hans D. Plug. (July 1985). "An Object within a Particle of Extraterrestrial Origin Compared with an Object of Presumed Terrestrial Origin" in *Astrophysics and Space Science*, vol. 113, 209–10 [doi:10.1023/A:1002432332393]; reprinted in *Astronomical Origins of Life: Steps towards Panspermia* (2000) edited by F. Hoyle and N. C. Wickramasinghe.

Hoyle, Fred with N. Chandra Wickramasinghe and M. K. Wallis. (1985c). "On the Nature of Dust Grains in the Comae of Comets Cernis and Bowell" in *Earth, Moon, and Planets*, vol. 33, 179–87.

Hoyle, Fred and R. A. Lyttleton. (July 1939). "Effect of Interstellar Matter on Climatic Variation" in *Mathematical Proceedings of the Cambridge Philosophical Society*, vol. 35, issue 3, 405–15 [doi:10.1017/s0305004100021150].

Hoyle, Fred and R. A. Lyttleton. (November 4, 1939). "The Evolution of the Stars" in *Mathematical Proceedings of the Cambridge Philosophical Society*, vol. 35, issue 4, 592–609.

Hoyle, Fred and R. A. Lyttleton. (June 1940). "On the Accretion of Interstellar Matter by Stars" in *Mathematical Proceedings of the Cambridge Philosophical Society*, vol. 36, issue 3, 325–30.

Hoyle, Fred and R. A. Lyttleton. (July 1940). "Note on Dr. Atkinson's Paper" in *Mathematical Proceedings of the Cambridge Philosophical Society*, vol. 36, issue 3, 323–24.

Hoyle, Fred and R. A. Lyttleton. (March 8, 1941). "On the Accretion Theory of Stellar Evolution" in *Monthly Notices of the Royal Astronomical Society*, vol. 102, issue 4, 226–36.

Hoyle, Fred and R. A. Lyttleton. (June 4, 1942). "On the Nature of Red Giant Stars" in *Monthly Notices of the Royal Astronomical Society*, vol. 102, issue 5, 218–25.

Hoyle, Fred with D. Noel F. Dunbar, William A. Wenzel, and Ward Whaling. (January 1953). "A State in C^{12} Predicted from Astrophysical Evidence" in *Physical Review*, vol. 92, 1095. (Klee 2002: 354; Kragh 2010: 34)

Hoyle, Fred and M. Schwarzchild. (June 1955). "On the Evolution of Type II Stars" in *The Astrophysical Journal Supplement*, vol. 2, 1 [doi:10.1086/190015].

Hoyle, Fred and Allan Sandage. (1956). "Second-Order Term in the Redshift-Magnitude Relation" in *Proceedings of the Astronomical Society of the Pacific*, vol. 68, issue 403, 301–7.

Hoyle, Fred with William A. Fowler, G. R. Burbidge, and E. M. Burbidge. (October 5, 1956). "Origin of the Elements in Stars" in *Science*, vol. 124, issue 3223, 611–14 [doi:10.1126/science.124.3223.611].

Hoyle, Fred and J. V. Narlikar. (June 13, 1961). "On the Counting of Radio Sources in the Steady-State Cosmology" in *Monthly Notices of the Royal Astronomical Society*, vol. 123, issue 2, 133–49.

Hoyle, Fred and John Elliot. (1962). *A for Andromeda: A Novel of Tomorrow.* London: Souvenir.

Hoyle, Fred and J. V. Narlikar. (July 5, 1962). "On the Counting of Radio Sources in the Steady-State Cosmology, II" in *Monthly Notices of the Royal Astronomical Society*, vol. 125, issue 1, 13–20.

Hoyle, Fred and J. V. Narlikar. (February 1962). "The Steady-State Model and the Ages of Galaxies" in *Observatory*, vol. 82, issue 926, 13–14.

Hoyle, Fred and William A. Fowler. (February 9, 1963). "Nature of Strong Radio Sources" in *Nature*, vol. 197, issue 4867, 533–35 [doi:10.1038/197533a0].

Hoyle, Fred and William A. Fowler. (1963). "On the Nature of Strong Radio Sources" in *Monthly Notices of the Royal Astronomical Society*, vol. 125, 169–76.

Hoyle, Fred and J. V. Narlikar. (January 7, 1964). "Time Symmetric Electrodynamics and the Arrow of Time in Cosmology" in *Proceedings of the Royal Society of London*, vol. 277, issue 1368, 1–23 [doi:1098/rspa.1964.0002].

Hoyle, Fred and R. J. Tayler. (September 12, 1964). "The Mystery of the Cosmic Helium Abundance" in *Nature*, vol. 203, issue 4950, 1108–10 [doi:10.1038/2031108a0].

Hoyle, Fred and J. V. Narlikar. (November 3, 1964). "A New Theory of Gravitation" in *Proceedings of the Royal Society of London*, vol. 282, issue 1389, 191–207. (Kragh 1996: 465)

Hoyle, Fred and others. (1965). *University of Denver Centennial Symposium: The Responsible Individual and a Free Society in an Expanding Universe.* Published for University of Denver by Big Mountain Press.

Hoyle, Fred and J. V. Narlikar. (1966 [March 1, 1965]). "A Radical Departure from the 'Steady-State' Concept in Cosmology" in *Proceedings of the Royal Society of London*, vol. 290, issue 1421, 162–76.

Hoyle, Fred and William A. Fowler. (January 28, 1967). "Gravitational Red-shifts in Quasi-stellar Objects" in *Nature*, vol. 213, issue 5074, 373–74 [doi:10.1038/213373a0].

Hoyle, Fred and G. R. Burbidge. (July 25, 1970). "The Log S-log z Diagram for Radio Galaxies and its Relation to Cosmology" in *Nature*, vol. 227, issue 5256, 359–61 [doi:10.1038/227359a0].

Hoyle, Fred and J. V. Narlikar. (January 1972a [July 27, 1971). "Cosmological Models in a Conformally Invariant Gravitational Theory—I: The Friedmann Models" in *Monthly Notices of the Royal Astronomical Society*, vol. 155, issue 3, 305–21.

Hoyle, Fred and J. V. Narlikar. (January 1972b [July 27, 1971). "Cosmological Models in a Conformally Invariant Gravitational Theory—II: A New Model" in *Monthly Notices of the Royal Astronomical Society*, vol. 155, issue 3, 323–35.

Hoyle, Fred and William A. Fowler. (February 9, 1973 [December 12, 1972]). "On the Origin of Deuterium" in *Nature*, vol. 241, issue 5389, 384–86 [doi:10.1038/241384a0].

Hoyle, Fred and Jayant V. Narlikar (1974). *Action at a Distance in Physics and Cosmology*. San Francisco: Freeman.

Hoyle, Fred and J. V. Narlikar. (1980). *The Physics-Astronomy Frontier*. San Francisco: Freeman. (Kragh 1996: 466)

Hoyle, Fred with Geoffrey Burbidge and Jayant V. Narlikar. (June 1993). "A Quasi-Steady State Cosmological Model with Creation of Matter" in *Astrophysical Journal*, part 1, vol. 410, issue 2, 437–57.

Hoyle, Fred with G. Burbidge and J. V. Narlikar. (1994 [November 24, 1993; June 10, 1993]). "Astrophysical Deductions from the Quasi-Steady-State Cosmology" in *Monthly Notices of the Royal Astronomical Society*, vol. 267, 1007–19.

Hoyle, Fred and Jayant V. Narlikar (January 1995 [December 31, 1994]). "Cosmology and Action-at-a-Distance Electrodynamics" in *Review of Modern Physics*, vol. 67, issue 1, 113–55 [doi:10.1103/RevModPhys.67.113].

Hoyle, Fred and Jayant V. Narlikar (1996). *Lectures on Cosmology and Action at a Distance Electrodynamics*. River Edge: World Scientific.

Hoyle, Fred with G. Burbidge and Jayant V. Narlikar. (1997 [October 28, 1996; June 21, 1996]). "On the Hubble Constant and the Cosmological Constant" in *Monthly Notices of the Royal Astronomical Society*, vol. 286, issue 1, 173–82 [doi:10.1093/mnras/286.1.173].

Hoyle, Fred with Geoffrey Burbidge and Jayant V. Narlikar. (2000). "The Observations Explained in Terms of the Quasi-Steady-State Model" (Chapter 16) in *A Different Approach to Cosmology: From a Static Universe through the Big Bang towards Reality*. Cambridge: Cambridge University Press.

Humboldt [Baron Friedrich Heinrich], Alexander von. (1866 [1845]). *Cosmos: A Sketch of a Physical Description of the Universe*, in four volumes, translation by Elise C. Otté. New York: Harper.

Hume, David. (1779). *Dialogues Concerning Natural Religion*. Reprinted in *David Hume: Principal Writings on Religion including Dialogues Concerning Natural Religion and the Natural History of Religion*, edited by J. C. A. Gaskin. Oxford: Oxford University Press, 1993.

Ibison, Michael and Stanley Jeffers. (1998). "A Double-Slit Diffraction Experiment to Investigate Claims of Consciousness-Related Anomalies" in *Journal of Scientific Exploration*, vol. 12, issue 4, 543–50.

Impey, Chris, editor. (2010). *Talking about Life: Conversations on Astrobiology*. Cambridge: Cambridge University Press.

Inge, William Ralph. (1933). *God and the Astronomers, Containing the Warburton Lectures*. London: Longmans, Green and Company.

Irvin, Jan with Andrew Rutajit and Jordon Maxwell. (2006). *Astrotheology and Shamanism: Unveiling the Law of Duality in Christianity and Other Religions*. San Diego: Book Tree.

Irvin, Jan and Andrew Rutajit. (2009). *Astrotheology and Shamanism: Christianity's Pagan Roots—A Revolutionary Reinterpretation of the Evidence*. Crestline: Gnostic Media.

Jacobs, Francine. (1983). *Cosmic Countdown: What Astronomers Have Learned about the Life of the Universe; Based on Works by Robert Jastrow*, foreword by Jastrow. New York: M. Evans.

Jaeger, Lydia. (2010). *Einstein, Polanyi, and the Laws of Nature*. West Conshohocken: Templeton.

Jain, Ravi with Maria C. Rivera, Jonathan E. Moore, and James A. Lake. (2003). "Horizontal Gene Transfer Accelerates Genome Innovation and Evolution" in *Molecular Biology and Evolution: Oxford Journals*, vol. 20, issue 10, 1598–1602.

Jaki, Stanley L. (1969). *The Paradox of Olbers' Paradox: A Case History of Scientific Thought*. New York: Herder.

Jaki, Stanley L. (1972). *The Milky Way: An Elusive Road to Science*. New York: Science History.

Jaki, Stanley L. (1974). *Science and Creation: From Eternal Cycles to an Oscillating Universe*. New York: Science History; revised 1986; Landham: University Press of America, 1990.

Jaki, Stanley L. (1978a]). *The Origin of Science and the Science of its Origin*. South Bend: Regnery/Gateway.

Jaki, Stanley L. (1978b). *The Road of Science and the Ways to God*. Chicago: University of Chicago Press.

Jaki, Stanley L. (1980). *Cosmos and Creator*. Edinburgh: Scottish Academic.

Jaki, Stanley L. (April 24, 1980). "Re-opening the Case for Natural Theology" [review of *Creation and the World of Science* (1979) by A. R. Peacocke] in *Nature*, vol. 284, issue 5758, 667–68 [doi:10.1038/284667a0].

Jaki, Stanley L. (1986). *Lord Gifford and His Lectures: A Centenary Retrospect*. Edinburgh: Scottish Academic.

Jaki, Stanley L. (1989). *God and the Cosmologists*. Washington: Regnery Gateway.

Jaki, Stanley L. (1990). *Cosmos in Transition: Studies in the History of Cosmology*. Tucson: Pachart.

James, William. (1878–1906). *Essays in Psychology* (Works of William James Series), introduction by W. Woodward. Cambridge: Harvard University Press.

James, William. (1890). *The Principles of Psychology*, two volumes. New York: Henry Holt.

James, William. (1892). *Psychology: Briefer Course*. New York: Henry Holt.

James, William. (1895). "Is Life Worth Living?" in *International Journal of Ethics*, vol. 6, issue 1.

James, William. (1897). *The Will to Believe and Other Essays in Popular Philosophy*. New York: Longmans, Green, and Company.

James, William. (1898). *Human Immortality: Two Supposed Objections to the Doctrine*. Boston: Houghton Mifflin.

James, William. (1899). *Talks to Teachers on Psychology: And to Students on Some of Life's Ideals*. New York: Henry Holt.

James, William. (1902). *The Varieties of Religious Experience: A Study in Human Nature* (Gifford Lectures, 1901–2). New York: Longmans, Green.

James, William. (1907). *Pragmatism: A New Name for Some Old Ways of Thinking.* New York, London: Longmans, Green; Cambridge: Harvard University Press, 1975.

James, William. (1909a). *The Meaning of Truth: A Sequel to "Pragmatism."* New York: Longmans, Green.

James, William. (1909b). *A Pluralistic Universe* (Hibbert Lectures). New York: Longmans, Green.

James, William. (1911a). *Memories and Studies.* New York: Longmans, Green.

James, William. (1911b). *Some Problems of Philosophy: A Beginning of an Introduction to Philosophy.* New York: Longmans, Green.

James, William. (1912). *Essays in Radical Empiricism*, edited by R. Perry. New York: Longmans, Green.

Jastrow, Robert. (1967). *Red Giants and White Dwarfs: The Evolution of Stars, Planets, and Life.* New York: Harper.

Jastrow, Robert. (1968). *Stars, Planets and Life: The Evolution of the Cosmos.* London: Heinemann.

Jastrow, Robert. (1969). *Space Science: A Guide Outlining Understandings, Fundamental Concepts, and Activities: Developed at Columbia University under the Auspices of the Director of the Summer Session in Cooperation with the Goddard Institute for Space Studies.* Washington: NASA.

Jastrow, Robert. (1971). *Red Giants and White Dwarfs: Man's Descent from the Stars.* New York: Harper.

Jastrow, Robert. (1977). *Until the Sun Dies.* New York: Norton; Warner, 1980.

Jastrow, Robert. (1978a). *God and the Astronomers.* New York: Norton; reprint 1980, 1992.

Jastrow, Robert. (June 25, 1978b). "In the Beginning Was the Bang—a Big One," in Sunday Opinion Section of the *Los Angeles Times*, K1, 6 [via ProQuest Historical].

Jastrow, Robert. (1979). *Red Giants and White Dwarfs.* New York: Warner; Norton, 1990.

Jastrow, Robert. (1981). *The Enchanted Loom: Mind in the Universe.* New York: Simon and Schuster.

Jastrow, Robert and Malcolm H. Thompson. (1972). *Astronomy: Fundamentals and Frontiers.* New York: Wiley; reprint 1977, 1974, 1984.

Jastrow, Robert and Michael Rampino. (2008). *Origins of Life in the Universe.* Cambridge: Cambridge University Press.

Jastrow, Robert, editor (1984). *The Essential Darwin.* Boston: Little Brown.

Jastrow, Robert and A. G. W. Cameron, editors. (1963). *Origin of the Solar System: Proceedings of a Conference Held at the Goddard Institute for Space Studies*, January 23–24, 1962, New York. New York: Academic.

Jeans, James Hopwood. (1927 [1925]). *The New Outlook in Cosmogony.* Cambridge: Cambridge University Press.

Jeans, James Hopwood. (1928). *Astronomy and Cosmogony*. Cambridge: Cambridge University Press.

Jeans, James Hopwood. (1931). *The Mysterious Universe*. New York: Macmillan.

Jeans, James Hopwood. (1937). *The Size and Age of the Universe*. Cambridge: Cambridge University Press.

Jeans, James Hopwood. (1944). *The Universe Around Us*. New York: Macmillan.

Jeans, James Hopwood. (1961). *Astronomy and Cosmogony*, preface by L. Motz. New York: Dover.

Jenkins, Alejandro and Gilad Perez. (January 2010). "Looking for Life in the Multiverse" in *Scientific American*, vol. 302, 42–49 [doi:10.1038/scientificamerican0110–42].

Jones-Bey, Hassaun Ali. (2015). "Wading in Troubled Waters: Optical, Musical, Cosmological, and Mathematical Resonances of an Opaque Theology." unpublished manuscript.

Joseph, Rhawn and N. Chandra Wickramasinghe. (September–October 2011). "Genetics Indicates Extraterrestrial Origins for Life: The First Gene—Did Life Begin Following the Big Bang?" in *Journal of Cosmology*, vol. 16, issue 21, 6832–61.

Just, Ernest Everett. (1939). *The Biology of the Cell Surface*. Philadelphia: P. Blakiston's Son and Company.

Kachappilly, Kurian. (2006). *Process: Implications and Applications*. Bangalore: Dharmaram.

Kagawa, Toyohiko. (2014). *Cosmic Purpose*, edited by T. Hastings. Eugene: Wipf and Stock.

Kaiser, David. (2011). *How the Hippies Saved Physics: Science, Counterculture, and the Quantum Revival*. New York: Norton.

Kaiser, R. I. with A. M. Stockton, Y. S. Kim, E. C. Jensen, and R. A. Mathies. (March 10, 2013 [October 10, 2012]). "On the Formation of Dipeptides in Interstellar Model Ices" in *The Astrophysical Journal*, vol. 765, issue 2, 111–20 [doi:10.1088/0004–637X/765/2/111].

Kak, Subhash, editor. (December 2009). "Cosmos, Quantum Physics, and Consciousness" in *Journal of Cosmology*, vol. 3.

Kanas, Nick (2009 [2007]). *Star Maps: History, Artistry and Cartography, corrected edition*. New York: Springer Science+Business Media.

Kane, Gordon. (December 16, 2011). "Particle Physics Is at a Turning Point" in *Nature News*, vol. 480, issue 7378, 415 [doi:10.1038/480415a].

Kanipe, Jeff. (2006). *Chasing Hubble's Shadows: The Search for Galaxies at the Edge of Time*. New York: Hill and Wang.

Kanipe, Jeff. (2009). *The Cosmic Connection: How Astronomical Events Impact Life on Earth*. Amherst: Prometheus.

Kanipe, Jeff and Dennis Webb. (2006). *Arp Atlas of Peculiar Galaxies: A Chronicle and Observer's Guide*. Richmond: Willmann-Bell.

Kant, Immanuel. (2008 [1755]). *Universal Natural History and Theory of the Heavens or An Essay on the Constitution of the Mechanical Origin of the*

Entire Structure of the Universe Based on Newtonian Principles [Allgemeine Naturgeschichte und Theorie des Himmels oder Versuch von der Verfassung und dem mechanischen Ursprunge des ganzen Weltgebäudes, nach Newtonischen Grundsätzen abgehandelt], translation by Ian Johnston. Arlington: Richer Resources.

Karim, L. M. with Fred Hoyle and N. C. Wickramasinghe (July 1983). "Interstellar Proteins and the Discovery of a New Absorption Feature at λ = 2800Å" in *Astrophysics and Space Science*, vol. 94, issue 1, 223–29.

Kaye, Howard L. (1986). *The Social Meaning of Modern Biology: From Social Darwinism to Sociobiology*. New Haven: Yale University Press; with new epilogue, New Brunswick: Transaction, 1997.

Keeling, Patrick J. and Jeffrey D. Palmer. (August 2008). "Horizontal Gene Transfer in Eukaryotic Evolution" in *Nature Reviews Genetics*, vol. 9, 605–18 [doi:10.1038/nrg2386].

Keller, Catherine and Anne Daniell, editors. (2002). *Process and Difference: Between Cosmological and Poststructuralist Postmodernisms*. Albany: State University of New York Press.

Keller, Catherine. (2003). *The Face of the Deep: A Theology of Becoming*. New York: Routledge.

Keller, Catherine. (2008). *On the Mystery: Discerning Divinity in Process*. Minneapolis: Fortress.

Kelley, David H. and Eugene F. Milone. (2011). *Exploring Ancient Skies: A Survey of Ancient and Cultural Astronomy*, second edition, foreword by A. F. Aveni. Dordrecht: Springer.

Kennedy, Gail. (1957). *Evolution and Religion: The Conflict between Science and Theology in Modern America*. Boston: Heath.

Keogh, Gary. (2014). *Reading Richard Dawkins: A Theological Dialogue with New Atheism*. Minneapolis: Fortress.

Kerferd, G. B. (1972 [1967]). "Anaxagoras of Clazomenae," *The Encyclopedia of Philosophy*, vol. 1, 115–17.

King, Martin Luther Jr. (2010 [1967]). *Where Do We Go From Here: Chaos or Community*, introduction by Vincent Harding. Boston: Beacon.

Kirshner, Robert P. (October 1994). "Earth's Elements" in *Scientific American*, vol. 271, issue 4, 58–65.

Kitagawa, Joseph M. and Charles H. Long, editors. (1969). *Myths and Symbols: Studies in Honor of Mircea Eliade*. Chicago: University of Chicago Press.

Klee, Robert. (2002). "The Revenge of Pythagoras: How a Mathematical Sharp Practice Undermines the Contemporary Design Argument in Astrophysical Cosmology" in *British Journal for the Philosophy of Science*, vol. 53, 331–54.

Klyce, Brig. (May 1996f). *Cosmic Ancestry* [online at www.panspermia.org/]. Astrobiology Research Trust.

Klyce, Brig. (August 2, 2009). "Panspermia and Horizontal Gene Transfer" in *Proceedings of SPIE – The International Society for Optical Engineering*, Instruments and Methods for Astrobiology and Planetary Missions XII conference vol. 7441 [doi:10.1117/12.832049].

Klyce, Brig and Chandra Wickramasinghe. (2003). "Creationism versus Darwinism: A Third Alternative" in *Darwinism, Design, and Public Education*, edited by John Angus Campbell and Stephen C. Meyers. East Lansing: Michigan State University, 543–48 [online at www.panspermia.org/thirdalt.htm].

Knight, David. (2013). "Chemical Sciences and Natural Theology" in *Oxford Handbook of Natural Theology*, edited by Russell Re Manning. Oxford: Oxford University Press.

Koning, Nico with S. Kwok, P. Bernath, A. Hjalmarson, and H. Olofsson. (2008). "Organic Molecules in the Spectral Line Survey of Orin KL with the Odin Satellite from 486–493 GHz and 541–577 GHz" in *Organic Matter in Space: Proceedings of the 251st Symposium of the International Astronomical Union*, February 18–22, Hong Kong, China, edited by Sun Kwok and Scott A. Sandford. Cambridge: Cambridge University Press.

Kragh, Helge. (1980). *On Science and Underdevelopment*. Roskilde Universitetsforiag.

Kragh, Helge. (1987). *An Introduction to the Historiography of Science*. Cambridge: Cambridge University Press.

Kragh, Helge. (1990). *Dirac: A Scientific Biography*. Cambridge: Cambridge University Press.

Kragh, Helge. (1993). "Big Bang Cosmology" and "Steady State Theory" in *Cosmology: Historical, Literary, Philosophical, Religious, and Scientific Perspectives*, edited by N. S. Hetherington. New York: Garland; Routledge, 2008.

Kragh, Helge. (1996). *Cosmology and Controversy: The Historical Development of Two Theories of the Universe*. Princeton: Princeton University Press.

Kragh, Helge. (1999). *Quantum Generations: A History of Physics in the Twentieth Century*. Princeton: Princeton University Press.

Kragh, Helge. (2004). *Matter and Spirit in the Universe: Scientific and Religious Preludes to Modern Cosmology*. London: Imperial College Press.

Kragh, Helge. (2007). *Conceptions of Cosmos: From Myths to the Accelerating Universe: A History of Cosmology*. Oxford: Oxford University Press.

Kragh, Helge. (2008). *Entropic Creation: Religious Contexts of Thermodynamics and Cosmology*. Burlington: Ashgate.

Kragh, Helge S. (2010). "When Is a Prediction Anthropic?: Fred Hoyle and the 7.65 MeV Carbon Resonance" in *PhilSci-Archive—An Archive for Preprints in Philosophy of Science*. University of Pittsburgh's University Library System, online at http://philsci-archive.pitt.edu/5332/1/3alphaphil.pdf.

Kragh, Helge. (November 2010). "An Anthropic Myth: Fred Hoyle's Carbon-12 Resonance Level" in *Archives for History of Exact Sciences*, vol. 64, issue 6, 721–51.

Kragh, Helge. (2011). *Higher Speculations: Grand Theories and Failed Revolutions in Physics and Cosmology*. Oxford: Oxford University Press.

Kragh, Helge S. (February–March 2011). "The Origin of the Modern Anthropic Principle" in *Journal of Cosmology*, vol. 13, issue 1, 3700–5.

Kragh, Helge. (2012). *Niels Bohr and the Quantum Atom: The Bohr Model of Atomic Structure, 1913–1925*. Oxford: Oxford University Press.

Kragh, Helge. (2012). "'The Wildest Speculation of All': Lemaître and the Primeval-Atom Universe" in *Georges Lemaître: Life, Science, and Legacy*, edited by R. D. Holder and S. Mitton. Berlin: Springer-Verlag.

Kragh, Helge. (2014). *Masters of the Universe: Conversations with Cosmologists of the Past*. Oxford: Oxford University Press.

Kragh, Helge and Robert W. Smith. (June 2003). "Who Discovered the Expanding Universe?" in *History of Science*, vol. 41, part 2, issue 132, 141–62.

Kragh, Helge S. and Dominique Lambert. (October 10, 2007). "Context of Discovery: Lemaître and the Origin of the Primeval-Atom Universe" in *Annals of Science*, vol. 64, issue 4, 445–70 [doi:10.1080/00033790701317692].

Kraus, Elizabeth M. (1979). *The Metaphysics of Experience: A Companion to Whitehead's Process and Reality*. Bronx: Fordham University Press.

Krause, Karl Christian Friedrich. (1829). *Vorlesungen über die Grundwahrheiten der Wissenshaft*. Germany: Göttingen. (Cooper 2006: 121)

Krause, Karl Christian Friedrich. (1900). *The Ideal of Humanity and Universal Federation*, translation by W. Hastie. Edinburgh: T & T Clark [originally *Das Urbild der Menschheit* (1812)]. (Cooper 2006: 121)

Krauss, Lawrence M. (2012). *A Universe from Nothing: Why There Is Something Rather than Nothing*, afterword by Richard Dawkins. New York: Free Press.

Krauss, Lawrence M. and Michael S. Turner. (September 2004). "A Cosmic Conundrum" in *Scientific American*, vol. 291, 70–77 [doi:10.1038/scientific-american0904-70].

Krauss, Lawrence M. and Michael S. Turner. (January 2006). "A Cosmic Conundrum" in *Scientific American*, vol. 15, 66–73 [doi:10.1038/scientific-american0206-66sp].

Krauss, Lawrence M. and Robert J. Scherrer. (March 2008). "End of Cosmology?" in *Scientific American*, vol. 298, 46–53 [doi:10.1038/scientificamerican0308-46].

Kuhn, Thomas S. (1957). *The Copernican Revolution: Planetary Astronomy in the Development of Western Thought*. Cambridge: Harvard University Press.

Kuhn, Thomas S. (1962). *The Structure of Scientific Revolutions*. Chicago: University of Chicago Press.

Kuhn, Thomas S. (1977). *The Essential Tension: Selected Studies in Scientific Tradition and Change*. Chicago: University of Chicago Press.

Kuhn, Thomas S. (1978). *Black-Body Theory and the Quantum Discontinuity, 1894–1912*. Oxford: Oxford University Press.

Kwok, Sun. (2001). *Cosmic Butterflies: The Colorful Mysteries of Planetary Nebulae*. Cambridge: Cambridge University Press.

Kwok, Sun. (2007). *Physics and Chemistry of the Interstellar Medium*. Sausalito: University Science.

Kwok, Sun. (2008). "Synthesis of Organic Compounds in the Late Stages of Stellar Evolution and Their Connection to the Solar System" in *Proceedings of the 10th Asian-Pacific Regional International Astronomical Union Meeting*, August 3–6, Kunming, China [and in Faculty of Science Conference Papers: HKU Scholars Hub, University of Hong Kong].

Kwok, Sun. (2009a). "Delivery of Complex Organic Compounds from Planetary Nebulae to the Solar System" in *International Journal of Astrobiology*, vol. 8, issue 3, 161–67 [doi:10.1017/S1473550409004492].

Kwok, Sun. (2009b). "Organic Matter in Space: From Star Dust to the Solar System" in *Astrophysics and Space Science*, vol. 319, issue 1, 5–21 [doi:10.1007/s10509–008–9965–6].

Kwok, Sun. (September–October 2011). "From 'Frontiers of Astronomy' to Astrobiology" in *Journal of Cosmology*, vol. 16, issue 13, 6643–60.

Kwok, Sun. (December [August 3] 2011). "Delivery of Complex Organic Compounds from Evolved Stars to the Solar System" in *Origins of Life and Evolution of Biospheres*, vol. 41, issue 6, 497–502 [doi:10.1007/s11084–011–9254–1].

Kwok, Sun. (2012). *Organic Matter in the Universe.* Weinheim: Wiley-VCH.

Kwok, Sun (2013). *Our Place in the Universe.* New Westminster: Angstrom.

Kwok, Sun and Scott A. Sandford, editors. (2008). *Organic Matter in Space: Proceedings of the 251st Symposium of the International Astronomical Union*, February 18–22, Hong Kong, China. Cambridge: Cambridge University Press.

Lal, Ashwini Kumar. (30 January 2010). "Big Bang? A Critical Review" in *Journal of Cosmology*, vol. 6, issue 11, 1533–47.

Landsburg, Steven E. (2009). *The Big Questions: Tackling the Problems of Philosophy with Ideas from Mathematics, Economics, and Physics.* New York: Free Press.

Langthaler, Rudolf and Kurt Appel. (2010). Dawkins' Gotteswahn: 15 Kritische Antworten auf Seine Atheistische Mission. Vienna: Böhlau.

Launius, Roger D, editor. (2012). *Exploring the Solar System: The History and Science of Planetary Exploration*. New York: Palgrave.

Lauretta, Dane S. and Harry Y. McSween Jr., editors. (2006). *Meteorites and the Early Solar System II.* Tucson: University of Arizona Press.

Leach, Javier. (2010). *Mathematics and Religion: Our Languages of Sign and Symbol.* West Conshohocken: Templeton.

Lee, Strobel and Jane Vogel. (2004). *The Case for a Creator: A Journalist Investigates Scientific Evidence That Points toward God.* Grand Rapids: Zondervan.

Lemaître, Georges Edouard. (1927). "Un Univers homogène de masse constant et de rayon croissant rendant compte de la vitesse radiale des nébuleuses extragalactiques" [A homogeneous Universe of constant mass and growing radius accounting for the radial velocity of extragalactic nebulae]. *Annales de la Société Scientifique de Bruxelles.*

Lemaître, Georges Edouard. (May 9, 1931). "The Beginning of the World from the Point of View of Quantum Theory" in *Nature*, vol. 127, 706 [doi:10.1038/127706a0].

Lemaître, Georges Edouard. (1946). *L'Hypothèse de l'Atome Primitive: Essai de Cosmogonie.* Neuchatel, Switzerland: Éditions du Griffon. (Kragh 1996: 469; O'Connor and Robertson 2008)

Lemaître, Georges Edouard. (1950). *The Primeval Atom: An Essay on Cosmogony*, translation by Betty H. and Serge A. Korff, preface by Ferdinand Gonseth, foreword by Henry Norris Russell. New York: D. Van Nostrand.

Lemaître, Georges Edouard and others. (1933). *Discussion sur l'évolution de l'univers*. Paris: Gauthier-Villars.
Lerner, Eric J. (1992 [1991]). *The Big Bang Never Happened*, paperback with new preface. New York: Random House.
Leslie, John, editor. (1990). *Physical Cosmology and Philosophy*. New York: Macmillan.
Leslie, John. (July 1994). "Anthropic Prediction" in *Philosophia*, vol. 23, issue 1–4, 117–44.
Leslie, John, editor. (1998). *Modern Cosmology and Philosophy*. Amherst: Prometheus.
Levere, Trevor Harvey. (1981). *Poetry Realized in Nature: Samuel Taylor Coleridge and Early Nineteenth-Century Science*. Cambridge: Cambridge University Press.
Levine, Stuart and Susan F. Levine. (2004). *Eureka / Edgar Allan Poe*. Urbana and Chicago: University of Illinois Press.
Lewis, Clarence Irving. (1918). *A Survey of Symbolic Logic*. Berkeley: University of California Press.
Lewis, Clarence Irving and Cooper Harold Langford. (1932). *Symbolic Logic*. New York: Century.
Liddle, Andrew. (2015). *An Introduction to Modern Cosmology*, third edition. Chichester: John Wiley; first edition 1999; second edition 2003.
Lieber, Lillian. (2008). *The Einstein Theory of Relativity: A Trip to the Fourth Dimension*. Philadelphia: Paul Dry.
Liebes, Sidney, Elisabet Sahtouris, and Brian Swimme. (1998). *A Walk Through Time: From Star Dust to Us*. New York: John Wiley.
Linde, A. D. (1982). "A New Inflationary Universe Scenario: A Possible Solution of the Horizon, Flatness, Homogeneity, Isotropy and Primordial Monopole Problems" in *Physics Letters B*, vol. 108, issue 6, 389–93 [doi:10.1016/0370–2693(82)91219–9].
Linde, A. D. (1983). "Chaotic Inflation" in *Physics Letters B*, vol. 129, issue 3–4, 177–81 [doi:10.1016/0370–2693(83)90837–7].
Livio, Mario. (2013). *Brilliant Blunders: From Darwin To Einstein—Colossal Mistakes by Great Scientists that Changed Our Understanding of Life and the Universe*. New York: Simon and Schuster.
Livio, M. with D. Hollowell, A. Weiss, and J. W. Truran. (July 27, 1989). "Anthropic Significance of the Existence of an Excited State of ^{12}C" in *Nature*, vol. 340, issue 6231, 281–84 [doi:10.1038/340281a0].
Lodge, Oliver. (1926). *Evolution and Creation*. New York: George H. Doran.
Long, Charles H. (1963). *Alpha: The Myths of Creation*. Chico: Scholars Press.
Long, Charles H. (1986). *Significations: Signs, Symbols, and Images in the Interpretation of Religion*. Philadelphia: Fortress.
Long, Charles H. (July 1987). *Myth and Mythology*. Washington: World & I [also www.worldandi.com or via Amazon Kindle].
Longair, Malcolm S. (2005). "Evolutionary Cosmologies: Then and Now" in *The Scientific Legacy of Fred Hoyle*, edited by D. Gough, foreword by M. Rees. Cambridge: Cambridge University Press, 2011 paperback.

Longair, Malcolm S. (2006). *The Cosmic Century: A History of Astrophysics and Cosmology*. Cambridge: Cambridge University Press; updated in 2013.
Longair, Malcolm S. (2011). *High Energy Astrophysics*, third edition. Cambridge: Cambridge University Press.
Loomer, Bernard M. (1971). "Whitehead's Method of Empirical Analysis" in *Process Theology: Basic Writings*, edited by E. Cousins. New York: Newman.
Lovejoy, Arthur O. (1936). *The Great Chain of Being: A Study of the History of an Idea* (William James Lectures at Harvard, 1933). Cambridge: Harvard University Press.
Lovell, Bernard (October 1964). "Radio Stars in the Galaxy" in *Observatory*, vol. 84, issue 942, 191–210.
Lowe, Victor. (1962). *Understanding Whitehead*. Baltimore: John Hopkins Press.
Lowe, Victor with Charles Hartshorne and A. H. Johnson. (1950). *Whitehead and the Modern World: Science, Metaphysics, and Civilization—Three Essays on the Thought of Alfred North Whitehead*, preface by A. C. Benjamin. Boston: Beacon.
Lyttleton, Raymond Arthur. (1953). *Comets and Their Origin*. Cambridge: Cambridge University Press.
Lyttleton, Raymond Arthur. (1961). *Man's View of the Universe*. London: M. Joseph.
Maasen, Sabine with Everett Mendelsohn and Peter Wingart, editors (1995). *Biology as Society, Society as Biology: Metaphors*. Dordrecht: Kluwer.
Macer, Darryl. (2015). "God, Life, Love, and Religions among Indigenous Peoples of the World" in *A Man of Many Parts: Essays in Honor of John Westerdale Bowker on the Occasion of His Eightieth Birthday*, edited by E. Lemcio. Eugene: Pickwick.
Maddox, John. (September 1, 1994). "Return of Cosmological Creation" in *Nature*, vol. 371, issue 6492, 11 [doi:10.1038/371011a0].
Maddox, John. (September 20, 2001). "Obituary: Fred Hoyle (1915–2001)" in *Nature*, vol. 413, issue 6853, 270 [doi:10.1038/35095162].
Maddox, John. (June 6, 2002). "Astronomy: The Hoyle Story" in *Nature*, vol. 417, issue 6889, 603–5 [doi:10.1038/417603a].
Maddox, Randy. (2009). "John Wesley's Precedent for Theological Engagement with the Natural Sciences" in *Divine Grace and Emerging Creation: Wesleyan Forays in Science and Theology of Creation*, edited by Thomas Jay Oord. Eugene: Pickwick Publications [also in *Wesleyan Theological Journal*, vol. 44, issue 1 (Spring 2009), 23–54, online at www.divinity.duke.edu/sites/default/files/documents/faculty-maddox/08_John_Wesley's_Precedent_with_ Science.pdf.
Malament, David B. (2002). *Reading Natural Philosophy: Essays in the History and Philosophy of Science and Mathematics*. Chicago: Open Court.
Malcolm, Norman. (January 1960). "Anselm's Ontological Arguments" in *Philosophical Review*, vol. 69, issue 1, 41–62.
Malone-France, Derek. (2012). "Process Metaphysics and Democratic Deliberation" in *Faith, Fallibility, and the Virtue of Anxiety: An Essay in Religion and Political Liberalism*. New York: Palgrave.

Mann, Charles C. (2013). *1493: Uncovering the New World Columbus Created.* New York: Knopf.

Manning, Russell Re, editor. (2013). *The Oxford Handbook of Natural Theology.* Oxford: Oxford University Press.

Manning, Russell Re and Michael Byrne, editors. (2013). *Science and Religion in the Twenty-First Century* (Boyle Lectures). London: SMC Press.

Manson, Neil A., editor. (2003). *God and Design: The Teleological Argument and Modern Science.* New York: Routledge.

Marcello, Barbieri. (1985). *The Semantic Theory of Evolution: Models of Scientific Thought*, vol. 2, foreword by René Thom, edited by Roger Hahn and René Thom. New York: Harwood.

Marcello, Barbieri. (2003). *The Organic Codes: An Introduction to Semantic Biology.* Cambridge: Cambridge University Press.

Margulis, Lynn and Dorion Sagan. (2002). *Acquiring Genomes: A Theory of the Origins of Species.* New York: Basic.

Margulis, Lynn and Eduardo Punset. (2007). *Mind, Life, and Universe: Conversations with Great Scientists of Our Time.* White River Junction: Chelsea Green.

Margulis, Lynn and Dorion Sagan. (2008). "The Role of Symbiogenesis in Evolution" in *Back to Darwin: A Richer Account of Evolution*, edited by John Cobb. Grand Rapids: Eerdmans.

Markham, Ian S. (2010). *Against Atheism: Why Dawkins, Hitchens, and Harris Are Fundamentally Wrong.* Malden: Wiley-Blackwell.

Martin, Ben R. (1976). "The Origins, Development and Capitulation of Steady-State Cosmology: A Sociological Study of Authority and Conflict in Science." University of Manchester MSc Thesis.

Martin, B. R. (February 1978). "Radio Astronomy Revisited: A Reassessment of the Role of Competition and Conflict in the Development of Radio Astronomy" in *The Sociological Review*, vol. 26, issue 1, 27–55 [doi:10.1111/j.1467–954X.1978.tb00123.x].

Mathis, J. S. with W. Rumpl and K. H. Nordsieck. (October 15, 1977). "Size Distribution of Interstellar Grains" in *The Astrophysical Journal*, Part 1, vol. 217, 425–33 [doi:10.1086/155591].

Matsuoka, K. with T. Nagao, R. Mailino, and others. (2011). "Chemical Properties in the Most Distant Radio Galaxy" in *Astronomy and Astrophysics*, vol. 532, issue L10.

Mautner, Michael N. (February–March 2010). "Seeding the Universe with Life: Securing Our Cosmological Future" in *Journal of Cosmology*, vol. 5, issue 26, 982–94 (also September 2010, vol. 11, issue 19).

Mayr, Ernst. (1988). *Toward a New Philosophy of Biology: Observations of an Evolutionist.* Belknap Press of Harvard University Press.

Mayr, Ernst. (1997). *This is Biology: The Science of the Living World.* Belknap Press of Harvard University Press.

McAleer, Neil. (2010). *Visionary: The Odyssey of Sir Arthur C. Clarke.* Baltimore: Clarke Project.

McCluskey, Stephen C. (1993). "Native American Cosmologies" in *Cosmology: Historical, Literary, Philosophical, Religious, and Scientific Perspectives*, edited by N. S. Hetherington. New York: Garland; Routledge, 2008.

McConnell, Craig. (October 2006). "The BBC, the Victoria Institute, and the Theological Context for the Big Bang—Steady State Debate" in *Science and Christian Belief*, vol. 18, issue 2, 151–68.

McCrea, William H. (1977). "Models, Laws, and the Universe" in *Cosmology, History, and Theology* [from Third International Colloquium at the University of Denver, November 5–8, 1974], edited by W. Yourgrau and A. D. Breck. New York: Plenum.

McCrea, William H. (1984). "The Influence of Radio Astronomy on Cosmology" in *The Early Years of Radio Astronomy: Reflections Fifty Years after Jansky*, edited by W. T. Sullivan. Cambridge: Cambridge University Press.

McFarland, Ian A. (2014). *From Nothing: A Theology of Creation*. Louisville: Westminster John Knox.

McFaul, Thomas R. and Al Brunsting. (2014). *God Is Here to Stay: Science, Evolution, and Belief in God*. Eugene: Wipf & Stock.

McGehee, Bobby. (2005). *New Universe Theory with the Laws of Physics*. Bloomington: AuthorHouse.

McGilchrist, Iain. (2010). *The Master and His Emissary: The Divided Brain and the Making of the Western World*. New Haven: Yale University Press.

McGrath, Alister. (2005). *Dawkins' God: Genes, Memes, and the Meaning of Life*. Malden: Wiley-Blackwell.

McGrath, Alister. (2006). *The Order of Things: Explorations in Scientific Theology*. Malden: Wiley-Blackwell.

McGrath, Alister E. (2010). *Science and Religion: An Introduction*, second edition. Malden: Wiley-Blackwell.

McGrath, Alister E. (2015). *Dawkins' God: Genes, Memes, Watchmakers, and Delusions*, second edition. Chichester: John Wiley.

McGrath, Alister E. and Joanna Collicutt McGrath. (2007). *Dawkins Delusion: Atheist Fundamentalism and the Denial of the Divine*. Downers Grove: InterVarsity.

McMullin, Ernan. (1981). "How Should Cosmology Relate to Theology" in *The Sciences and Theology in the Twentieth Century*, edited by A. R. Peacocke. Notre Dame: University of Notre Dame Press.

McMullin, Ernan. (1992). *The Social Dimensions of Science*. Notre Dame: University of Notre Dame Press.

McMullin, Ernan, editor. (1985). *Evolution and Creation*. Notre Dame: University of Notre Dame Press.

McMullin, Ernan, editor. (2005). *The Church and Galileo*. Notre Dame: University of Notre Dame Press.

McNaughton, N. J. and C. T. Pillinger. (December 11, 1980). "Comets and the Origin of Life" in *Nature*, vol. 288, issue 5791, 540 [doi:10.1038/288540a0].

McSween, Harry Y. Jr. and Gary R. Huss. (2010). *Cosmochemistry*. Cambridge: Cambridge University Press.

Mesle, C. Robert. (2008). "The Future of Process-Relational Thought" in *Process-Relational Philosophy: An Introduction to Alfred North Whitehead*. West Conshohocken: Templeton.

Metcalfe, N. with T. Shanks, A. Campos, R. Fong, and J. P. Gardner. (September 19 [August 13], 1996). "Galaxy Formation at High Redshifts" in *Nature*, vol. 383, issue 6597, 236–39 [doi:10.1038/383236a0].

Meyer, William J. (2010). *Metaphysics and the Future of Theology: The Voice of Theology in Public Life*. Eugene: Pickwick.

Middleton, Eric. (2007). *New Flatlanders: A Seeker's Guide to the Theory of Everything*. Philadelphia: Templeton.

Miller, Kenneth R. (1999). *Finding Darwin's God: A Scientist's Search for Common Ground between God and Evolution*. New York: Cliff Street.

Miller, Ron. (2014). *Recentering the Universe: The Radical Theories of Copernicus, Kepler, and Galileo*. Minneapolis: Twenty-First Century Books.

Miller, Stanley L. (May 15, 1953). "A Production of Amino Acids under Possible Primitive Earth Conditions." *Science*, New Series, vol. 117, issue 3046, 528–29.

Miller, Stanley L. and Harold C. Urey. (July 1959). "Organic Compound Synthesis on the Primitive Earth." *Science*, vol. 130, issue 3370, 245–51.

Miller, Stanley L. and Leslie E. Orgel. (1974). *The Origins of Life on the Earth*. Upper Saddle: Prentice Hall.

Misner, Charles W. (February 1968 [June 22, 1967]). "Isotropy of the Universe" in *The Astrophysical Journal*, vol. 151, 431–57 [doi:10.1086/149448].

Misner, Charles W. (1977). "Cosmology and Theology" in *Cosmology, History, and Theology* [from Third International Colloquium at the University of Denver, November 5–8, 1974], edited by W. Yourgrau and A. D. Breck. New York: Plenum.

Mitton, Simon. (2005). *Conflict in the Cosmos: Fred Hoyle's Life in Science*, foreword by Owen Gingerich. Washington: Joseph Henry.

Mitton, Simon. (2005). *Fred Hoyle: A Life in Science*. London: Aurum; Cambridge: Cambridge University Press.

Mitton, Simon. (2008). "Hoyle, Fred" in *New Dictionary of Scientific Biography*. Detroit: Scribner's.

Moltmann, Jürgen. (1985). *God in Creation: A New Theology of Creation and the Spirit of God* (Gifford Lectures 1984–85). San Francisco: Harper.

Moltmann, Jürgen. (2009). "Sighs, Signs, and Significance: Natural Science and a Hermeneutics of Nature" in *Divine Grace and Emerging Creation: Wesleyan Forays in Science and Theology of Creation*, edited by Thomas Jay Oord. Eugene: Pickwick.

Moltmann, Jürgen. (2011). "Is the World Unfinished? On Interactions between Science and Theology in the Concepts of Nature, Time, and the Future" in *Theology*, vol. 114, issue 6, 403–13.

Monteiro-Ferreira, Ana. (2014). *Demise of the Inhuman: Afrocentricity, Modernism, and Postmodernism*, foreword by Molefi Kete Asante. Albany: State University of New York Press.

Moore, James R. (1979). *The Post-Darwinian Controversies: A Study of the Protestant Struggle to Come to Terms with Darwin in Great Britain and America, 1870–1900.* Cambridge: Cambridge University Press.

Morbidelli, A. with J. Chambers, J. I. Lunine, J. M. Petit, F. Robert, G. B. Valsecchi, and K. E. Cyr. (November 2000). "Source Regions and Timescales for the Delivery of Water to the Earth" in *Meteoritics and Planetary Science*, vol. 35, issue 6, 1309–20 [doi:10.1111/j.1945–5100.2000.tb01518.x].

Morris, Bertram. (1941). "The Art-process and the Aesthetic Fact in Whitehead's Philosophy" in *The Philosophy of Alfred North Whitehead, The Library of Living Philosophers, Vol. 3*, edited by Paul Arthur Schilpp. La Salle: Open Court.

Morris, Simon Conway, editor. (2008). *The Deep Structure of Biology: Is Convergence Sufficiently Ubiquitous to Give a Directional Signal?* West Conshohocken: Templeton.

Morrison, Reg. (2008). "Hydrogen: Humanity's Maker and Breaker" in *Back to Darwin: A Richer Account of Evolution*, edited by John B. Cobb Jr. Grand Rapids: Eerdmans.

Moseley, Henry. (1847). *Astro-Theology.* London: A. Varnham.

Motta, V. with E. Mediavilla, J. A. Muñoz, E. Falco, C. S. Kochanek, S. Arribas, B. García-Lorenzo, A. Oscoz, and M. Serra-Ricart. (2002). "Detection of the 2175 Å Extinction Feature at $z = 0.83$" in *The Astrophysical Journal*, vol. 574, issue 2, 719–25 [doi:10.1086/341118].

Munitz, Milton K. (April 1952). "Scientific Method in Cosmology" in *Philosophy of Science*, vol. 19, issue 2, 108–30.

Munitz, Milton K. (1954). "Creation and the 'New' Cosmology" in *British Journal for the Philosophy of Science*, vol. 5, issue 17, 32–46 [doi:1093/bjps/V.17.32].

Munitz, Milton K. (1957a). *Space, Time and Creation: Philosophical Aspects of Scientific Cosmology.* Glencoe: Free Press. (Kragh 1996: 474)

Munitz, Milton K. (1957b). *Theories of the Universe from Babylonian Myth to Modern Science.* Glencoe: Free Press.

Munitz, Milton K. (1974). *The Mystery of Existence: An Essay in Philosophical Cosmology.* New York: New York University Press.

Munitz, Milton K. (1986). *Cosmic Understanding: Philosophy and Science of the Universe.* Princeton: Princeton University Press.

Muray, Leslie A. (1988). *An Introduction to the Process Understanding of Science, Society, and the Self.* Lewiston: Edwin Mellen.

Murdin, Paul. (2009). *Secrets of the Universe: How We Discovered the Cosmos.* Chicago: University of Chicago Press.

Murdoch, Iris. (1992). *Metaphysics as a Guide to Morals* [based on 1982 Gifford Lectures]. London: Chatto and Windus.

Murphy, Nancey C. (1990). *Theology in the Age of Scientific Reasoning.* Ithaca: Cornell University Press.

Murphy, Nancey C. (1994). *Reasoning and Rhetoric in Religion.* Valley Forge: Trinity.

Murphy, Nancey C. (1996). *Beyond Liberalism and Fundamentalism: How Modern and Postmodern Philosophy Set the Theological Agenda*. Valley Forge: Trinity.

Murphy, Nancey C. (1997). *Anglo-American Postmodernity: Philosophical Perspectives on Science, Religion, and Ethics*. Boulder: Westview.

Murphy, Nancey C. (2006). *Bodies and Souls, or Spirited Bodies?* Cambridge: Cambridge University Press.

Murphy, Nancey C. (2011). "Cosmopolis: How Astronomy Affects Philosophies of Human Nature and Religion" in *Astronomy and Civilization in the New Enlightenment: Passions of the Skies [Analecta Husserliana: The Yearbook of Phenomenological Research, Volume CVII]*, edited by Anna-Teresa Tymieniecka and Attila Grandpierre. Dordrecht: Springer.

Murphy, Nancey C. and George F. R. Ellis. (1996). *On the Moral Nature of the Universe: Theology, Cosmology, and Ethics*. Minneapolis: Fortress.

Murphy, Nancey C. and Warren S. Brown. (2007). *Did My Neurons Make Me Do It?: Philosophical and Neurobiological Perspectives on Moral Responsibility and Free Will*. Oxford: Oxford University Press.

Murphy, Nancey C. and William R. Stoeger. (2007). *Evolution and Emergence: Systems, Organisms, Persons*. Oxford: Oxford University Press.

Murphy, Nancey C. with Robert John Russell and William R. Stoeger, S. J., editors. (2007). *Physics and Cosmology: Scientific Perspectives on the Problem of Natural Evil*. Vatican City State: Vatican Observatory Press.

Murphy, Nancey C. with George F. R. Ellis and Timothy O'Connor, editors. (2009). *Downward Causation and the Neurobiology of Free Will*. Berlin: Springer-Verlag.

Murphy, Nancey C. and Christopher C. Knight, editors. (2010). *Human Identity at the Intersection of Science, Technology, and Religion*. Burlington: Ashgate.

Mustol, John Stephen. (2012). *Dusty Earthlings: Living as Eco-Physical Beings in God's Eco-Physical World*, foreword by Nancey C. Murphy. Eugene: Cascade.

Nagel, Thomas. (2012). *Mind and Cosmos: Why the Materialist Neo-Darwinian Conception of Nature Is Almost Certainly False*. Oxford: Oxford University Press.

Nagy, Bartholomew with George Claus and Douglas J. Hennessy. (March 24, 1962). "Organic Particles Embedded in Minerals in the Orgueil and Ivuna Carbonaceous Chondrites" in *Nature*, vol. 193, issue 4821, 1129–33 [doi:10.1038/1931129a0].

Nagy, Bartholomew with Kurt Fredriksson, Harold C. Urey, George Claus, Christian A. Anderson, and others. (April 13, 1963). "Electron Probe Microanalysis of Organized Elements in the Orgueil Meteorite" in *Nature*, vol. 198, issue 4876, 121–25 [doi:10.1038/198121a0].

Napier, W. M. (January 30, 2004). "A Mechanism for Interstellar Panspermia" in *Monthly Notices of the Royal Astronomical Society*, vol. 348, issue 1, 46–51.

Napier, W. M. with J. T. Wickramasinghe and N. C. Wickramasinghe. (August 12, 2004). "Extreme Albedo Comets and the Impact Hazard" in *Monthly Notices of the Royal Astronomical Society*, vol. 355, 191–95 [doi:10.1111/j.1365–2966.2004.08309.x].

Narlikar, Jayant V. (1973). "Steady State Defended" in *Cosmology Now*, edited by L. John. London: BBC. (Kragh 1996: 467, 474)

Narlikar, Jayant V. (September 1992). "Concepts of 'Beginning' and 'Creation' in Cosmology" in *Philosophy of Science*, vol. 59, issue 3, 361–71.

Narlikar, Jayant V. (September 1992). "Concepts of 'Beginning' and 'Creation' in Cosmology" in *Philosophy of Science*, vol. 59, issue 3, 361–71.

Narlikar, Jayant V. (2005). "Alternative Ideas in Cosmology" in *The Scientific Legacy of Fred Hoyle*, edited by D. Gough, foreword by M. Rees. Cambridge: Cambridge University Press, 2011 paperback.

Narlikar, Jayant V. and N. C. Wickramasinghe. (October 7, 1967). "Microwave Background in a Steady State" in *Nature*, vol. 216, issue 5110, 43–44 [doi:10.1038/216043a0].

Narlikar, Jayant V. and N. C. Wickramasinghe. (March 30, 1968). "Interpretation of Cosmic Microwave Background" in *Nature*, vol. 217, issue 5135, 1235–36 [d oi:10.1038/2171235a0].

Narlikar, Jayant V. with Indu Banga and Chhanda Gupta, editors. (1992). *Philosophy of Science: Perspectives from Natural and Social Sciences*. Shimla: Indian Institute of Advanced Study; Delhi: Munshiram Manoharlal.

Narlikar, Jayant V. and Geoffrey Burbidge. (2008). "An Alternative Cosmology" in *Facts and Speculations in Cosmology*. Cambridge: Cambridge University Press.

Needham, Joseph. (1936). *Order and Life*. New Haven: Yale University Press.

Needham, Joseph. (1941). "A Biologist's View of Whitehead's Philosophy" in *The Philosophy of Alfred North Whitehead, The Library of Living Philosophers, Vol. 3*, edited by Paul Arthur Schilpp. La Salle: Open Court.

Needham, Joseph. (1947). *Science and Society in Ancient China*. London: Watts.

Needham, Joseph. (1954). *Science and Civilisation in China*. Cambridge: Cambridge University Press.

Needham, Joseph. (1969). *Grand Titration: Science and Society in East and West*. London: Allen and Unwin.

Needham, Joseph. (1979). *History and Human Values: A Chinese Perspective for World Science and Technology: Inaugural Martin Wickramasinghe Lecture, 1978*. Nawala: Martin Wickramasinghe Trust Publisher.

Needham, Joseph and Walter Pagel, editors. (1975 [1938]). *Background to Modern Science*. New York: Arno.

Needham, Joseph and Colin Ronan. (1993). "Chinese Cosmology" in *Cosmology: Historical, Literary, Philosophical, Religious, and Scientific Perspectives*, edited by N. S. Hetherington. New York: Garland; Routledge, 2008.

Neville, Robert C. (1968). *God the Creator: On the Transcendence and Presence of God*, drawings by Beth Neville. Chicago: University of Chicago Press.

Neville, Robert C. (1974). *The Cosmology of Freedom*. New Haven: Yale University Press.

Neville, Robert C. (1995). *Creativity and God: A Challenge to Process Theology*, new edition. Albany: State University of New York Press [originally, Seabury 1980].

Neville, Robert C. (2013). *Ultimates: Philosophical Theology*. Albany: State University of New York Press.

Nicolescu, Basarab. (1996). *La Transdisciplinarité: Manifesto*. Monaco: Editions du Rocher.

Nicolescu, Basarab. (2002). *Manifesto of Transdisciplinarity*, translation by Karen-Claire Voss. Albany: State University of New York Press.

Nicolescu, Basarab. (2013). *From Modernity to Cosmodernity: Science, Culture, and Spirituality*. Albany: State University of New York Press.

Nicolescu, Basarab, editor. (2008). *Transdisciplinarity: Theory and Practice*. Cresskill: Hampton.

Northrop, Filmer S. C. (1941). "Whitehead's Philosophy of Science" in *The Philosophy of Alfred North Whitehead, The Library of Living Philosophers, Vol. 3*, edited by P. A. Schilpp. La Salle: Open Court.

Noterdaeme, P. with C. Ledoux, R. Srianand, P. Petitjean, and S. Lopex. (2009). "Diffuse Molecular Gas at High Redshift—Detection of CO Molecules and the 2175 Å Dust Feature at $z = 1.64$" in *Astronomy and Astrophysics*, vol. 503, 765–70 [ArXiv:0906.1585v1].

Nye, Bill. (2014). *Undeniable: Evolution and the Science of Creation*, edited by C. S. Powell. New York: St. Martin's.

Nyíri, Kristóf. (2013). "Images in Natural Theology" in *The Oxford Handbook of Natural Theology*, edited by Russell Re Manning. Oxford: Oxford University Press.

O'Connor, J. J. and E. F. Robertson. (July 2008). "Georges Henri-Joseph-Edouard Lemaître." MacTutor History of Mathematics at www-history.mcs.st-andrews.ac.uk/Biographies/Lemaitre.html [accessed December 20, 2012].

Ogden, Schubert M. (1966). *The Reality of God and Other Essays*. New York: Harper.

Ogden, Schubert M. (April 1968). "God and Philosophy: A Discussion with Antony Flew" in *Journal of Religion*, vol. 48, issue 2, 161–81.

Ogden, Schubert M. (Spring 1975). "The Criterion of Metaphysical Truth and the Senses of Metaphysics" in *Process Studies*, vol. 5, issue 1, 47–48.

Ogden, Schubert M. (1975). "'Theology and Falsification' in Retrospect: A Reply" in *The Logic of God: Theology and Verification*, edited by Malcolm L. Diamond and Thomas V. Litzenburg Jr. Indianapolis: Bobbs-Merrill.

Ogden, Schubert M. (1979). *Faith and Freedom: Toward a Theology of Liberation*. Nashville: Abingdon; revised and enlarged, Eugene: Wipf & Stock, 2005.

Ogden, Schubert M. (1983). "Myth" in *The Westminster Dictionary of Christian Theology*, edited by A. Richardson and J. Bowden. Philadelphia: Westminster.

Ogden, Schubert M. (1984a). "The Experience of God: Critical Reflections on Hartshorne's Theory of Analogy" in *Existence and Actuality: Conversations with Charles Hartshorne*, edited by John Cobb and Franklin Gamwell. Chicago: University of Chicago Press.

Ogden, Schubert M. (Spring 1984b). "Process Theology and the Wesleyan Witness" in *Perkins School of Theology Journal*, vol. 37, issue 3, 18–33 [reprinted in *Thy Nature and Thy Name Is Love: Wesleyan and Process Theologies in Dialogue* (2001), edited by Bryan P. Stone and Thomas Jay Oord].

Ogden, Schubert M. (Summer 1985). "The Metaphysics of Faith and Justice" in *Process Studies*, vol. 14, issue 2, 87–101.

Ogden, Schubert M. (1986). "What Is Theology" in *On Theology*. New York: Harper.

Ogden, Schubert M. (2008). "Theology without Metaphysics?" in *Religion and the End of Metaphysics*, edited by Dewi Z. Phillips and Mario von der Ruhr. Tubingen: Mohr Siebeck.

O'Leary, Margaret R. (2008). *Anaxagoras and the Origin of Panspermia Theory*. Bloomington: iUniverse.

Olupona, Jacob K., editor. (2004). *Beyond Primitivism: Indigenous Religious Traditions and Modernity*. New York: Routledge.

Oord, Thomas Jay. (2009). *Creation Made Free: Open Theology Engaging Science*. Eugene: Pickwick.

Oord, Thomas Jay. (2010). *Defining Love: A Philosophical, Scientific, and Theological Engagement*. Grand Rapids: Brazos.

Oord, Thomas Jay. (2014). *Theologies of Creation: Creatio Ex Nihilo and its Rivals*. New York: Routledge.

Oord, Thomas Jay, editor. (2009). *Divine Grace and Emerging Creation: Wesleyan Forays in Science and Theology of Creation*. Eugene: Pickwick.

Oort, J. H. (January 1950). "The Structure of the Cloud of Comets Surrounding the Solar System and a Hypothesis Concerning Its Origin" in *Bulletin of the Astronomical Institutes of the Netherlands*, vol. 11, 91–110.

Oparin, Aleksandr Ivanovičh. (1938). *The Origin of Life*, translation with annotations by Sergius Morgulis. New York: Macmillan; Dover 1953, 2003.

Orgel, Leslie E. (October 1994). "The Origin of Life on the Earth" in *Scientific American*, vol. 271, issue 4, 76–83.

Ormerod, Neil. (2015). *A Public God: Natural Theology Reconsidered*. Minneapolis: Fortress.

Orr, David W. (1992). *Ecological Literacy: Education and the Transition to a Postmodern World*. Albany: State University of New York Press.

Ostovich, Steven T. (1990). *Reason in History: Theology and Science as Community Activities*. Atlanta: Scholars.

Otto, Helen Tzima. (1995). *Their Earthshaking News: When Science and Theology Speak the Same Urgent Language*. Houston: Verenikia.

Otto, Marc and Michael Lodahl. (2009). "Mystery and Humility in John Wesley's Narrative Ecology" in *Divine Grace and Emerging Creation: Wesleyan Forays in Science and Theology of Creation*, edited by Thomas Oord. Eugene: Pickwick.

Page, Ruth. (2004). "Panentheism and Pansyntheism: God in Relation" in *In Whom We Live and Move and Have Our Being: Panentheistic Reflections on God's Presence in a Scientific World*, edited by P. Clayton and A. Peacocke. Grand Rapids: Eerdmans.

Pagel, Bernard Ephraim Julius. (1997). *Nucleosynthesis and Chemical Evolution of Galaxies*. Cambridge: Cambridge University Press; 2009 second edition.

Paley, William. (1802). *Natural Theology; or, Evidences of the Existence and Attributes of the Deity; Collected from the Appearances of Nature.* London: Printed for J. Faulder.

Pannenberg, Wolfhart. (1990). *Metaphysics and the Idea of God*, translation by Philip Clayton. Grand Rapids: Eerdmans.

Pannenberg, Wolfhart. (November 2007). *The History of Nature: Essays on Science and Theology*, edited by Niels Henrik Gregersen. West Conshohocken: Templeton.

Pasachoff, Jay M. (August 31, 2006). "In Retrospect: Out of the Darkness" [review of *Black Cloud* (1957) by Fred Hoyle] in *Nature*, vol. 442, issue 7106, 986 [doi:10.1038/442986a].

Pasachoff, Jay M. and William A. Fowler. (May 1974). "Deuterium in the Universe" in *Scientific American*, vol. 230, 108–19 [doi:10.1038/scientificamerican0574–108].

Pasteur, Louis. (February 6, 1860). "Expériences relatives aux générations spontanées" in *Comptes rendus de l'Académie des Sciences*, vol. 50, 303–7.

Pasteur, Louis. (May 7, 1860a). "De l'origines des ferments: Nouvelle experiences relatives aux generations dites spontanées" in *Comptes rendus de l'Académie des Sciences*, vol. 50, 849–54.

Pasteur, Louis. (September 3, 1860b). "Nouvelle experiences relatives aux generations dites spontanées" in *Comptes rendus de l'Académie des Sciences*, vol. 51, 348–53.

Pasteur, Louis. (November 5, 1860c). "Suite à uneprécédente communication relative aux generations dites spontanées" in *Comptes rendus de l'Académie des Sciences*, vol. 51, 675–78.

Pasteur, Louis. (1861). "Mémoiresur les corpuscules organisés qui existent dans l'atmosphère: Examen de la doctrine des generations spontanées" in *Annales des Sciences Naturelles* (partie Zoologique), 4e série, vol. 16, 5–68.

Patrizi, Fancesco. (1591). *Nova de Universis Philosophia in qua aristotelica methodo non per motum sed per lucem et lumina ad primam causam ascenditur. Deinde propria Patricii methodo tota in contemplationem venit Divinitas. Postremo, methodo platonica, rerum universitas a conditore Deo deducitur*, Ferrara: Mammarelli, reprinted in Venice in 1593, London in 1611. (Purnell 22 November 2004).

Paul, Iain. (1986). *Science and Theology in Einstein's Perspective.* Edinburgh: Scottish Academic.

Paulson, Steve. (2010). *Atoms and Eden: Conversations on Religion and Science.* Oxford: Oxford University Press.

Peacocke, Author Robert. (1979). *Creation and the World of Science.* Oxford: Oxford University Press.

Peacocke, Author Robert (1986). *God and the New Biology.* San Francisco: Harper.

Peacocke, Author Robert (1993 [1990]). *Theology for a Scientific Age: Being and Becoming—Natural, Divine, and Human.* Minneapolis: Fortress.

Peacocke, Author Robert (2004 [1979]). *Creation and the World of Science: The Re-shaping of Belief.* Oxford: Oxford University Press.

Peacocke, Author Robert. (2004). *Evolution: The Disguised Friend of Faith?: Selected Essays*. Philadelphia: Templeton.

Peacocke, Author Robert (2005). *The Palace of Glory: God's World and Science*. Hindmarsh: Australian Theological Forum Press.

Peacocke, Author Robert (2007). *All That Is: A Naturalistic Faith for the Twenty-First Century—A Theological Proposal with Responses from Leading Thinkers in the Religion-Science Dialogue*, edited by P. Clayton. Minneapolis: Fortress.

Peacocke, Author Robert, editor. (1985). *Reductionism in Academic Disciplines*. Guildford: Society for Research into Higher Education and Nelson Publishing.

Peacocke, Author Robert, editor (1981). *The Sciences and Theology in the Twentieth Century*. Notre Dame: University of Notre Dame Press.

Peacock, John A. (1999). *Cosmological Physics*. Oxford: Oxford University Press.

Peck, A. L., translator. (1970). *Aristotle: Historia Animalium, Books IV-VI*. Cambridge: Harvard University Press.

Pecker, Jean-Claude and Jayant Narlikar, editors. (2006). *Current Issues in Cosmology*. Cambridge: Cambridge University Press.

Pedraja, Luis. (2002). "Whitehead, Deconstruction, and Postmodernism" in *Process and Difference: Between Cosmological and Poststructuralist Postmodernisms*, edited by Catherine Keller and Anne Daniell. Albany: State University of New York Press.

Pedraja, Luis. (2011). "Alfred North Whitehead" in *Beyond the Pale: Reading Theology from the Margins*, edited by Miguel A. De La Torre and Stacey M. Floyd-Thomas. Louisville: Westminster John Knox.

Peebles, P. James with David N. Schramm, Edwin L. Turner, and Richard G. Kron. (October 1994). "The Evolution of the Universe" in *Scientific American*, vol. 271, issue 4, 53–57.

Penrose, Roger. (1991 [1989]). *The Emperor's New Mind: Concerning Computers, Minds, and the Laws of Physics*, foreword by Martin Gardner. New York: Penguin.

Penrose, Roger. (1994). *Shadows of the Mind: A Search for the Missing Science of Consciousness*. Oxford: Oxford University Press.

Penrose, Roger. (2004). "Speculative Theories of the Early Universe" and "The Anthropic Principle" in *Road to Reality: A Complete Guide to the Laws of the Universe*. London: Jonathan Cape.

Penrose, Roger. (2010). "Conformal Cyclic Cosmology" in *Cycles of Time: An Extraordinary New View of the Universe*. London: Bodley Head; New York: Knopf, 2011.

Penrose, Roger. (2011). *Collected Works*. Oxford: Oxford University Press.

Penrose, Roger and Wolfgang Rindler. (1984, 1986). *Spinors and Space-Time, Volumes 1 and 2*. Cambridge: Cambridge University Press.

Penrose, Roger and others. (1999 [1997]). *The Large, the Small, and the Human Mind*, new and revised, edited by Malcolm Longair. Cambridge: Cambridge University Press.

Penrose, Roger and Stuart Hameroff, editors. (April–May 2011). "Consciousness and the Universe" in *Journal of Cosmology*, vol. 14.

Penzias, Arno A. (1977). "An Observational View of the Cosmos" in *Cosmology, History, and Theology* [from the Third International Colloquium at University of Denver, November 5–8, 1974], edited by W. Yourgrau and A. D. Breck. New York: Plenum.

Penzias, Arno A. and Robert W. Wilson. (July 1965). "A Measurement of Excess Antenna Temperature at 4080 Mc/s" in *The Astrophysical Journal Letters*, vol. 142, 419–21 [doi:10.1086/148307].

Peters, Eugene H. (1970). *Hartshorne and Neoclassical Metaphysics: An Interpretation*. Lincoln: University of Nebraska Press.

Peters, Ted, editor. (1989). *Cosmos as Creation: Theology and Science in Consonance*. Nashville: Abingdon.

Peters, Ted. (1998). *Science and Theology: The New Consonance*. Boulder: Westview.

Peters, Ted. (2013). "Astrotheology" in *The Routledge Companion to Modern Christian Thought*, edited by Chad Meister and James Beilby. New York: Routledge, 838–53.

Peters, Ted. (June 2014). "Astrotheology: A Constructive Proposal" in *Zygon*, vol. 449, issue 2, 443–47 [doi:10.1111/zygo.12094].

Peters, Ted and Gaymon Bennett, editors (2003). *Bridging Science and Religion*. Minneapolis: Fortress.

Peters, Ted and Martinez Hewlett. (2003). *Evolution from Creation to New Creation: Conflict, Conversation, and Convergence*. Nashville: Abingdon.

Peters, Ted and Martinez Hewlett. (2006). *Can You Believe in God and Evolution: A Guide for the Perplexed*. Nashville: Abingdon.

Pflug, Hans Dieter. (November 26, 1981). "Extraterrestrial Life: New Evidence of Microfossils in the Murchison Meterorite." A public lecture in Cardiff.

Pflug, Hans Dieter. (1984). "Ultrafine Structure of the Organic Matter in Meteorites" in *Fundamental Studies and the Future of Science*, edited by Chandra Wickramasinghe. Cardiff: University College Cardiff Press.

Pflug, Hans Dieter and H. Haescheke-Boyer. (August 9, 1979). "Combined Structural and Chemical Analysis of 3,800-Myr-Old Microfossils" in *Nature*, vol. 280, issue 5722, 483–86 [doi:10.1038/280483a0].

Poe, Edgar Allan. (1848). *Eureka: An Essay on the Material and Spiritual Universe*. Also titled "Eureka: A Prose Poem" in *Eureka / Edgar Allan Poe*, edited with introduction, notes, and text variants by Stuart Levine and Susan F. Levine. Champaign: University of Illinois Press, 2004.

Poe, Harry Lee. (October 30, 2009). "Eureka: Edgar Allan Poe's Big Bang Theory and the Power of Imagination." Lecture sponsored by the Metanexus Institute.

Polkinghorne, John C. (c. 1986). *One World: The Interaction of Science and Theology*. Princeton: Princeton University Press.

Polkinghorne, John C. (c. 1988). *Science and Creation: The Search for Understanding*. Boston: New Science Library Press.

Polkinghorne, John C. (1989). *Science and Providence: God's Interaction with the World*. Boston: New Science; Philadelphia: Templeton, 2005.

Polkinghorne, John C. (1991). *Reason and Reality: The Relationship between Science and Theology*. Philadelphia: Trinity.

Polkinghorne, John C. (1994). *The Faith of a Physicist: Reflections of a Bottom-Up Thinker* (Gifford Lectures). Princeton: Princeton University Press.

Polkinghorne, John C. (1998a). *Belief in God in an Age of Science*. New Haven: Yale University Press.

Polkinghorne, John C. (1998b). *Science and Theology: An Introduction*. Minneapolis: Fortress.

Polkinghorne, John C. (2002). *Traffic in Truth: Exchanges between Science and Theology*. Minneapolis: Fortress.

Polkinghorne, John C. (2010a). *The Polkinghorne Reader: Science, Faith and the Search for Meaning*, edited by Thomas Jay Oord. West Conshohocken: SPCK/Templeton.

Polkinghorne, John C. (2013). "Mathematics and Natural Theology" in *The Oxford Handbook of Natural Theology*, edited by Russell Re Manning. Oxford: Oxford University Press.

Polkinghorne, John C., editor. (2010b). *Trinity and an Entangled World: Relationality in Physical Science and Theology*. Grand Rapids: Eerdmans.

Polkinghorne, John C. and Michael Welker, editors (2000). *The End of the World and the Ends of God: Science and Theology on Eschatology*. Harrisburg: Trinity.

Pomerleau, Wayne P. (2014). "William James (1842–1910)" in *Internet Encyclopedia of Philosophy: A Peer-Reviewed Academic Resource*, online at www.iep.utm.edu/james-o/#H1.

Ponnamperuma, Cyril and Ruth Mack. (May 28, 1965). "Nucleotide Synthesis under Possible Primitive Earth Conditions" in *Science*, vol. 148, issue 3674, 1221–23.

Ponnamperuma, Cyril, editor. (c. 1981). *Comets and the Origin of Life: Proceedings of the Fifth College Park Colloquium on Chemical Evolution*, October 29–31, 1980, University of Maryland. Boston: Kluwer.

Popper, Karl R. (1935 [1934]). *Logik der Forschung: Zur Erkenntnistheorie der Modernen Naturwissenschaft, von Karl Popper*. Vienna: Springer.

Popper, Karl R. (1959). *The Logic of Scientific Discovery*. New York: Basic.

Popper, Karl R. (1982). *The Open Universe: An Argument for Indeterminism*. Totowa: Rowman and Littlefield.

Popper, Karl R. (1982). *Quantum Theory and the Schism in Physics*. London: Unwin Hyman.

Popper, Karl R. (1983). *Realism and the Aim of Science*. Totowa: Rowman and Littlefield.

Popper, Karl R. and John C. Eccles. (1977). *The Self and Its Brain*. Berlin: Springer-Verlag.

Primack, Joel R. and Nancy Ellen Abrams. (2006). *The View from the Center of the Universe: Discovering Our Place in the Cosmos*. New York: Riverhead.

Procter, James D., editor. (2009). *Envisioning Nature, Science, and Religion*. West Conshohocken: Templeton.

Przybylski, A. (April 2, 1966). "Abundance Analysis of the Peculiar Star HD 101065" in *Nature*, vol. 210, issue 5031, 20–22 [doi:10.1038/210020a0].

Purnell, Fred. (November 22, 2004). "Fancesco Patrizi" in *Stanford Encyclopedia of Philosophy*, edited by E. N. Zalta, online at http://plato.stanford.edu/entries/patrizi/#Bib [accessed June 29, 2014].

Quine, Willard V. (1941). "Whitehead and the Rise of Modern Logic" in *The Philosophy of Alfred North Whitehead, The Library of Living Philosophers, Vol. 3*, edited by Paul Arthur Schilpp. La Salle: Open Court.

Rabkin, Eric S. (May 10, 2012). "Defining Science Fiction." Lecture 13 from the lecture series Masterpieces of the Imaginative Mind: Literature's Most Fantastic Works [accessed via YouTube, January 9, 2014].

Radhakrishnan, V. (July 25, 1993). "Jan Oort and Radio Astronomy" in *Current Science*, vol. 65, 134–37.

Radin, Dean with Leena Michel and others. (2012). "Consciousness and the Double-Slit Interference Pattern: Six Experiments" in *Physics Essays*, vol. 25, issue 2, 157–71.

Rae, Murray with Hilary Regan and John Stenhouse, assisted by A. Wood and J. Aitken, editors. (1994). *Science and Theology: Questions at the Interface*. Grand Rapids: Eerdmans.

Rauf, Kani and Chandra Wickramasinghe. (January 2010). "Evidence for Biodegradation Products in the Interstellar Medium" in *International Journal of Astrobiology*, vol. 9, issue 1, 29–34.

Ravetz, J. R. (1981). "Varieties of Scientific Experience" in *The Sciences and Theology in the Twentieth Century*, edited by A. R. Peacocke. Notre Dame: University of Notre Dame Press.

Raymo, Chat. (February 2005). "Big Bang vs. Steady State: How the Big Bang Theory Won the 20th Century's Biggest Cosmological Debate" [review of *Big Bang: The Origin of the Universe* (2005) by Simon Singh] in *Scientific American*, vol. 292, 98–100 [doi:10.1038/scientificamerican0205–98].

Reagan, Michael, editor. (2002). *Inside the Mind of God: Images and Words of Inner Space*, introduction by Sharon Begley. Philadelphia: Templeton.

Rees, M. J. and J. P. Ostriker. (June 1977 [November 5, 1976; July 7, 1976]). "Cooling, Dynamics and Fragmentation of Massive Gas Clouds: Clues to the Masses and Radii of Galaxies and Clusters" in *Monthly Notices of the Royal Astronomical Society*, vol. 179, 541–59.

Rees, Martin. (1997). *Before the Beginning: Our Universe and Others*, foreword by Stephen Hawking. Cambridge: Helix Books.

Rees, Martin. (2001). *Our Cosmic Habitat*. Princeton: Princeton University Press.

Rees, Martin (2005). Foreword to *The Scientific Legacy of Fred Hoyle*, edited by D. Gough. Cambridge: Cambridge University Press, 2005, 2011 paperback.

Rees, Martin. (January 10, 2011). "Life in the Cosmos." Madingley Lecture at University of Cambridge.

Regan, Hilary D. and Mark Wm. Worthing, editors. (2002). *Interdisciplinary Perspectives on Cosmology and Biological Evolution*. Adelaide: Australian Theological Forum.

Reid, Jennifer I. (2008). *Religion and Global Culture: New Terrain in the Study of Religion and the Work of Charles H. Long*. Landham: Lexington Books.

Rescher, Nicholas. (1996). *Process Metaphysics: An Introduction to Process Philosophy*. Albany: State University of New York Press.

Ridley, Mark and Alan Grafen, editors. (2006). *Richard Dawkins: How a Scientist Changed the Way We Think: Reflections by Scientists, Writers, and Philosophers*. Oxford: Oxford University Press.

Roberts, J. Deotis. (1991). *A Philosophical Introduction to Theology*. Philadelphia: Trinity.

Roe, Nicholas, editor. (2001). *Samuel Taylor Coleridge and the Sciences of Life*. Oxford: Oxford University Press.

Rolston III, Holmes. (1988). *Environmental Ethics: Duties to and Values in the Natural World*. Philadelphia: Temple University Press.

Rolston III, Holmes. (2006). *Science and Religion: A Critical Survey*. Philadelphia: Temple University Press.

Rolston III, Holmes. (2010). *Three Big Bangs: Matter-Energy, Life, Mind*. New York: Columbia University Press.

Rolston III, Holmes, editor. (1995). *Biology, Ethics, and the Origins of Life*. Boston: Jones and Bartlett.

Romanes, George J. (February 15, 1983). "Natural Selection and Natural Theology" in *Nature*, vol. 27, 362–64 [doi:10.1038/027362a0].

Romanes, George J. (April 5, 1983). "Natural Selection and Natural Theology" in *Nature*, vol. 27, 528–29 [doi:10.1038/027528a0].

Rood, R. T. and A. Renzini, editors. (1997). *Advances in Stellar Evolution: Proceedings of the Workshop Stellar Ecology*, June 23–29, 1996, Marciana Marina, Elba, Italy. Cambridge: Cambridge University Press.

Rosen, Steven M. (c2008). *The Self-Evolving Cosmos: A Phenomenological Approach to Nature's Unity-in-Diversity*. Hackensack: World Scientific.

Rosenberg, Alex and Robert Arp, editors. (2010). *Philosophy of Biology: An Anthology*. Chichester: Wiley-Blackwell.

Rowan-Robinson, Michael. (December 22, 1972). "Steady State Obituary?" in *Nature*, vol. 240, issue 5382, 439 [doi:1038/240439a0].

Rubenstein, Mary-Jane (2008). *Strange Wonder: The Closure of Metaphysics and the Opening of Awe*. New York: Columbia University Press.

Rubenstein, Mary-Jane. (June 2012). "Cosmic Singularities: On the Nothing and the Sovereign" in *Journal of the American Academy of Religion*, vol. 80, issue 2, 485–517.

Rubenstein, Mary-Jane. (2014). *Worlds without End: The Many Lives of the Multiverse... in which are discussed pre-, early-, and post-, modern cosmologies; the sundry arguments for and against them; the striking peculiarities of their adherents and detractors; the shifting boundaries of science, philosophy, and religion; and the stubbornly persistent question of whether creation has been "designed."* New York: Columbia University Press.

Rucker, Rudy. (1983 [1982]). *Infinity and the Mind: The Science and Philosophy of the Infinite*. New York: Bantam.

Rudnicki, Konrad. (June 1991). *The Cosmologist's Second: The Riddle of Time in Theories of the Universe*, translation by M. Lipson. Hudson: Lindisfarne.

Rudwick, Martin. (1981). "Senses of the Natural World and Senses of God: Another Look at the Historical Relation of Science and Religion" in *Sciences and Theology in the Twentieth Century*, edited by A. R. Peacocke. Notre Dame: University of Notre Dame Press.

Rushton, Simon K. and Rob Gray. (October 5, 2006). "Hoyle's Observations Were Right on the Ball" in *Nature*, vol. 443, issue 7111, 506 [doi:10.1038/443506d].

Russell, Bertrand. (1935). *Religion and Science*. London: Butterworth.

Russell, Bertrand. (1948). "Whitehead and Principia Mathematica" in *Mind*, vol. 57, issue 226, 137–38.

Russell, Peter. (April 2003). *From Science to God: A Physicist's Journey into the Mystery of Consciousness*. Novato: New World.

Russell, Robert John with Nancey Murphy and C. J. Isham, editors. (1993). *Quantum Cosmology and the Law of Nature: Scientific Perspectives on Divine Action*. Vatican City State: Vatican Observatory Publications.

Russell, Robert John with Nancey Murphy and Arthur R. Peacocke, editors. (1995). *Chaos and Complexity: Scientific Perspectives on Divine Action*. Vatican City State: Vatican Observatory Publications.

Ryan, Frank P. (2004). "Human Endogenous Retroviruses in Health and Disease: A Symbiotic Perspective" in *Journal of the Royal Society of Medicine*, vol. 97, issue 12, 560–65.

Ryle, Martin. (August 1955). "Radio Stars and Their Cosmological Significance" in *Observatory*, vol. 75, issue 887, 137–47.

Ryle, Martin. (June 3, 1961). "Radio Astronomy and Cosmology" in *Nature*, vol. 190, issue 4779, 852–54 [doi:10.1038/190852a0].

Ryle, Martin and R. W. Clarke (January 30, 1961). "An Examination of the Steady-State Model in the Light of Some Recent Observations of Radio Sources" in *Monthly Notices of the Royal Astronomical Society*, vol. 122, issue 4, 349–62.

Sagan, Carl. (1977). *The Dragons of Eden: Speculations on the Evolution of Human Intelligence*. New York: Random House.

Sagan, Carl. (1979). *Broca's Brain: Reflections on the Romance of Science*. New York: Random House.

Sagan, Carl. (1980). *Cosmos*. New York: Random House; Ballantine, 1985.

Sagan, Carl. (1985). *Contact: A Novel*. New York: Simon & Schuster.

Sagan, Carl. (October 1994a). "The Search for Extraterrestrial Life" in *Scientific American*, vol. 271, issue 4, 92–99.

Sagan, Carl. (1994b). *Pale Blue Dot: A Vision of the Human Future in Space*. New York: Random House.

Sagan, Carl. (1995). *The Demon-Haunted World: Science as a Candle in the Dark*. New York: Random House.

Sagan, Carl. (1997). *Billions and Billions: Thoughts on Life and Death at the Brink of the Millennium*. New York: Random House.

Sagan, Carl. (2006). *The Varieties of Scientific Experience: A Personal View of the Search for God*, edited by Ann Druyan. New York: Penguin [originally, "Search for Who We Are" (unpublished Gifford Lectures), 1985].

Sagan, Carl and B. N. Khare. (1971). "Long-wavelength Ultraviolet Photo Production of Amino Acids on the Primitive Earth" in *Science*, vol. 173, 417–20 [doi:10.1126/science.173.3995.417].

Sagan, Carl and Ann Druyan. (1985). *Comet*. New York: Random House; Ballantine, 1997.

Sagan, Dorion. (2008). "Evolution, Complexity, and Energy Flow" in *Back to Darwin: A Richer Account of Evolution*, edited by John Cobb. Grand Rapids: Eerdmans.

Sakata, A. with N. Nakagawa, T. Iguchi, S. Isobe, M. Morimoto, F. Hoyle, and N. C. Wickramasinghe. (March 17, 1977). "Spectroscopic Evidence for Interstellar Grain Clumps in Meteoritic Inclusions" in *Nature*, vol. 266, issue 5599, 241 [doi:10.1038/266241].

Salpeter, E. E. and N. C. Wickramasinghe. (May 3, 1969). "Alignment of Interstellar Grains by Cosmic Rays" in *Nature*, vol. 222, issue 5192, 442–44 [doi:10.1038/222442a0].

Sandford, Scott A. (2008). "Organics in the Samples Returned from Comet 81P/Wild 2 by the Stardust Spacecraft" in *Organic Matter in Space: Proceedings of the 251st Symposium of the International Astronomical Union*, February 18–22, Hong Kong, China, edited by Sun Kwok and Scott A. Sandford. Cambridge: Cambridge University Press.

Sandford, Scott with Louis Allamandola and others. (2009). "Understanding the Evolution of Molecules in Space: Implications for Astrophysics, Astrochemistry, and Astrobiology" in *Astro2010: The Astronomy and Astrophysics Decadal Survey, Science White Papers*, no. 257 [2009astro2010S.257S].

Sargent, Wallace L. W. (2005). "Fred Hoyle's Major Work in the Context of Astronomy and Astrophysics Today" in *The Scientific Legacy of Fred Hoyle*, edited by D. Gough, foreword by M. Rees. Cambridge: Cambridge University Press, 2011 paperback.

Sarmah, B. P. with S. K. Banerjee, S. V. Dhurandhar, and J. V. Narlikar. (June 2006 [May 5, 2006]). "On Searches for Gravitational Waves from Mini-Creation Events by Laser Interferometric Detectors" in *Monthly Notices of the Royal Astronomical Society*, vol. 369, issue 1, 89–96 [doi:10.1aaa/j.1365–2966.2006.10262.x].

Savage, B. D. and J. S. Mathis. (1979). "Observed Properties of Interstellar Dust" in *Annual Review of Astronomy and Astrophysics*, vol. 17, 73–111 [doi:10.1146/annurev.aa.17.090179.000445].

Scalzi, Giuliano with Laura Selbmann and others. (June 12, 2012 [December 5, 2011]). "LIFE Experiment: Isolation of Cryptoendolitic Organisms from Antarctic Colonized Sandstone Exposed to Space and Simulated Mars Conditions on the International Space Station" in *Origins of Life and Evolution of Biospheres*, vol. 42, issue 2–3, 253–62 [doi:10.1007/s11084–012–9282–5].

Scharf, Caleb. (January 26, 2012). "Cosmology: Plucked from the Vacuum" in *Nature*, vol. 481, issue 7382, 440 [doi:10.1038/481440a].

Scharf, Caleb A. (2014). *The Copernicus Complex: Our Cosmic Significance in a Universe of Planets and Probabilities.* New York: Scientific American/Farrar, Straus and Giraux.

Schenk, Jim, editor. (c2006). *What Does God Look Like in an Expanding Universe?* Cincinnati: ImagoEarth.

Schild, Rudolf with R. Gabriel Joseph, Chandra Wickramasinghe, Robert J. C. McLean, Richard B. Hoover, Gilbert V. Levin, Milton Wainwright, and Max K. Wallis. (2014 [2010]). *Biological Cosmology, Astrobiology, Extraterrestrial Life.* Cosmology Science Publishers.

Schild, Rudolph E. (June 2012). "Introduction to Astro-Theology" in *Journal of Cosmology*, vol. 19, issue 1, 8547–51.

Schilpp, Paul Arthur. (1941). "Whitehead's Moral Philosophy" in *The Philosophy of Alfred North Whitehead, The Library of Living Philosophers, Vol. 3*, edited by Paul Arthur Schilpp. La Salle: Open Court.

Schwarzachild, Martin. (1958). *Structure and Evolution of the Stars.* Princeton: Princeton University Press; New York: Dover, 1965.

Sciama, Dennis William. (1959). *The Unity of the Universe*, with diagrams by J. Wedge. Garden City: Doubleday; Mineola: Dover, 2009.

Scott, Eugenie C. (2009). *Evolution vs. Creationism: An Introduction*, second edition. Berkeley: University of California Press.

Scott, Wilbur S., Introducer. (2002). *Complete Tales and Poems of Edgar Allan Poe.* Edison: Castle.

Seckbach, Joseph, editor. (2012). *Genesis—in the Beginning: Precursors of Life, Chemical Models and Early Biological Evolution.* Dordrecht: Springer.

Selin, Helaine, editor with advisory editor Sun Xiaochun. (2000). *Astronomy across Cultures: The History of Non-Western Astronomy.* Dordrecht: Kluwer.

Sellars, Roy Wood. (1941). "Philosophy of Organism and Physical Realism" in *The Philosophy of Alfred North Whitehead, The Library of Living Philosophers, Vol. 3*, edited by Paul Arthur Schilpp. La Salle: Open Court.

Shapiro, Robert (May 13, 1993). "Life, the Universe and Anything Goes" [review of *Our Place in the Cosmos* (1993) by Hoyle and Wickramasinghe] in *Nature*, vol. 363, issue 6425, 124 [doi:10.1038/363124a0].

Shaviv, Giora. (2012). *The Synthesis of the Elements: The Astrophysical Quest for Nucleosynthesis and What It Can Tell Us about the Universe.* Heidelberg: Springer-Verlag.

Shaw, Jane. (2015). "The Moral Imagination and a Sense of God" in *A Man of Many Parts: Essays in Honor of John Westerdale Bowker on the Occasion of his Eightieth Birthday*, edited by E. E. Lemcio. Eugene: Pickwick.

Sheldrake, Rupert. (c. 2012). *Science Set Free: Ten Paths to New Discovery.* New York: Deepak Chopra Books.

Sherburne, Donald W. (1970). *A Whiteheadian Aesthetic; Some Implications of Whitehead's Metaphysical Speculation*, foreword by F. S. C. Northrop. Hamden: Archon.

Sherburne, Donald W., editor. (1981 [c. 1966]). *A Key to Whitehead's Process and Reality*. Chicago: University of Chicago Press.

Sherburne, Donald W. and Janusz A. Polanowske, editors. (2004). *Whitehead's Philosophy: Points of Connection*. Albany: State University of New York Press.

Shields, George W. (May 2001). "Physicalist Panexperientialism and the Mind-Body Problem" in *American Journal of Theology and Philosophy*, vol. 22, issue 2, 133–54.

Shivaji, Sisinthy with Sreenivas Ara, Snajay Kumar Singh, Sunil Bandi, Aditya Singh, and Anil Kumar Pinnaka. (December 2012). "Draft Genome Sequence of Strain B3W22, Isolated from the Upper Atmosphere" in *Journal of Biology*, vol. 194, issue 23, 6624–25 [doi:10.1128/JB.01651–12].

Shivaji, Sisinthy with Preeti Chaturvedi, Zareena Begum, Pavan Kumar Pindi, R. Manorama, D. Ananth Padmanaban, Yogesh S. Shouche, Shrikant Pawar, Parag Vaishampayan, C. B. S. Dutt, G. N. Datta, R. K. Manchanda, U. R. Rao, P. M. Bhargava, and J. V. Narlikar. (2009). "*Janibacter hoylei* sp. nov., *Bacillus isronensis* sp. nov. and *Bacillus aryabhattai* sp. nov., Isolated from Cryotubes Used for Collecting Air from the Upper Atmosphere" in *International Journal of Systematic and Evolutionary Microbiology*, vol. 59, issue 12, 2977–86 [doi:10.1099/ijs.0.002527–0].

Shklovskiĭ, I. S. and Carl Sagan. (1966). *Intelligent Life in the Universe*, translation by Paula Fern. San Francisco: Holden-Day.

Shults, F. LeRun with Nancey Murphy and Robert John Russell, editors. (2009). *Philosophy, Science and Divine Action*. Leiden: Brill.

Sia, Marian F. (2010). *From Question to Quest: Literary-Philosophical Enquiries into the Challenges of Life*. Newcastle upon Tyne: Cambridge Scholars.

Sia, Santiago. (1985). *God in Process Thought: A Study in Charles Hartshorne's Concept of God*. Dordrecht: Martinus Nijhoff.

Sia, Santiago. (2004). *Religion, Reason and God: Essays in the Philosophies of Charles Hartshorne and A. N. Whitehead*. New York: Peter Lang.

Sia, Santiago. (2010). "Literary Insights, Philosophical Development and Ethical Reflections: An Exploration into their Interconnectedness" in *Ethical Contexts and Theoretical Issues: Essays in Ethical Thinking*. Newcastle upon Tyne: Cambridge Scholars.

Sia, Santiago and André Cloots. (1999). *Framing a Vision of the World: Essays in Philosophy, Religion and Science*. Leuven: Leuven University Press.

Silk, Joseph. (February 1968). "Cosmic Black-Body Radiation and Galaxy Formation" in *The Astrophysical Journal*, vol. 152, 459–71 [doi:10.1086/149449].

Silk, Joseph. (1980). *The Big Bang: The Creation and Evolution of the Universe*, foreword by Dennis Sciama. San Francisco: Freeman; second edition 1989; third edition 2001.

Silk, Joseph (February 11, 1982). "Variations on Cosmology and Astrophysics" in *Nature*, vol. 295, issue 5849, 461 [doi:10.1038/295461a0].

Silk, Joseph. (1994a). *Cosmic Enigmas*. Woodbury: AIP Press.

Silk, Joseph. (1994b). *A Short History of the Universe*. New York: Freeman.

Silk, Joseph. (2005). *On the Shores of the Unknown: A Short History of the Universe*. Cambridge: Cambridge University Press.

Silk, Joseph. (2008 [2006]). *Infinite Cosmos: Questions from the Frontiers of Cosmology*. Oxford: Oxford University Press.

Silk, Joseph. (2009). *Horizons of Cosmology: Exploring Worlds Seen and Unseen*. West Conshohocken: Templeton.

Silk, Joseph. (March 8, 2014). "What Can We Do When Cosmology Raises Questions it Cannot Answer? Call in the Philosophers?" in *New Scientist*, vol. 221, 26–28 [arxiv.org/ftp/arxiv/papers/1404/1404.3128.pdf].

Singh, Simon. (2004). *Big Bang: The Origin of the Universe*. New York: Fourth Estate; San Francisco: HarperCollins, 2005.

Skrbina, David. (March 2003). "Panpsychism as an Underlying Theme in Western Philosophy: A Survey Paper" in *Journal of Consciousness Studies*, vol. 10, issue 3.

Skrbina, David. (April 8, 2007). "Panpsychism" in *Internet Encyclopedia of Philosophy*, online at www.iep.utm.edu/panpsych/.

Smart, John Jamieson Carswell. (1989). *Our Place in the Universe: A Metaphysical Discussion*. Oxford: Blackwell.

Smart, Ninian. (1972 [1967]). "Rāmānuja" in *The Encyclopedia of Philosophy*, vol. 7, 64–65.

Smith, Tony. (1991). "Agricultural Science and Ideology" in *The Role of Ethics in Social Theory: Essays from a Habermasian Perspective*. Albany: State University of New York Press.

Smolin, Lee. (1997). *The Life of the Cosmos*. Oxford: Oxford University Press.

Smolin, Lee. (2002). *Three Roads to Quantum Gravity*. New York: Basic.

Smolin, Lee. (2006). *The Trouble with Physics: The Rise of String Theory, the Fall of a Science, and What Comes Next*. Boston: Houghton Mifflin.

Smolin, Lee. (2007 [July 29, 2004]). "Scientific Alternatives to the Anthropic Principle" in *Universe or Multiverse*, edited by Bernard Carr. Cambridge: Cambridge University Press [arXiv:hep-th/0407213v3].

Smolin, Lee. (October 28, 2010). "Cosmology: Space-time Turn Around" in *Nature*, vol. 467, issue 7319, 1034–35 [doi:10.1038/4671034a].

Smolin, Lee. (2013). *Time Reborn: From the Crisis in Physics to the Future of the Universe*. Boston: Houghton Mifflin.

Snyder, Laura J. (2011). *The Philosophical Breakfast Club: Four Remarkable Friends* [William Whewell, Charles Babbage, John Herschel, and Richard Jones] *Who Transformed Science and Changed the World*. New York: Crown.

Southgate, Christopher. (2008). *The Groaning of Creation: God, Evolution, and the Problem of Evil*. Louisville: Westminster John Knox.

Southgate, Christopher. (2013). "Natural Theology and Ecology" in *The Oxford Handbook of Natural Theology*, edited by Russell Re Manning. Oxford: Oxford University Press.

Stannard, Russell. (1982). *Science and the Renewal of Belief*. London: SCM Press; Philadelphia: Templeton, 2004.

Stannard, Russell. (2000). *The God Experiment: Can Science Prove the Existence of God?* Mahwah: Hidden Spring.
Stannard, Russell (2012). *Science and Belief: The Big Issues.* London: A Lion Book.
Stannard, Russell, editor. (2000). *God for the 21st Century.* Philadelphia: Templeton.
Stapp, Henry P. (2007). *Mindful Universe: Quantum Mechanics and the Participating Observer.* Heidelberg: Springer-Verlag.
Starkman, Glenn D. and Roberto Trotta. (November 16 [July 17], 2006). "Why Anthropic Reasoning Cannot Predict Lambda" in *Physical Review Letters*, vol. 97, issue 201301 [doi:10.1103/PhysRevLett.97.201301].
Starobinsky, A. A. (1980). "A New Type of Isotropic Cosmological Models Without Singularity" in *Physics Letters B*, vol. 91, issue 1, 99–102 [doi:10.1016 /0370–2693(80)90670-X].
Staune, Jean, editor. (2006). *Science and the Search for Meaning: Perspectives from International Scientists.* Philadelphia: Templeton.
Steinhardt, Paul J. (March 22, 2011). "The Inflation Debate" in *Scientific American*, vol. 304, 36–43 [doi:10.1038/scientificamerican0411–36].
Stenger, Victor J. (2007). *God: The Failed Hypothesis: How Science Shows that God Does Not Exist.* Amherst: Prometheus.
Stenger, Victor J. (2009). *The New Atheism: Taking a Stand for Science and Reason.* Amherst: Prometheus.
Stenger, Victor J. (2011). *The Fallacy of Fine-Tuning: Why the Universe Is Not Designed for Us.* Amherst: Prometheus.
Stenger, Victor J. (2012). *God and the Folly of Faith: The Incompatibility of Science and Religion.* Amherst: Prometheus.
Stenger, Victor J. (2013). *God and the Atom.* Amherst: Prometheus.
Stenger, Victor J. (2014). *God and the Multiverse: Humanity's Expanding View of the Cosmos.* Amherst: Prometheus.
Stevens, Clifford J. (1969). *Astrotheology for the Cosmic Adventure.* Techny: Divine Word.
Stewart, Melville Y., editor. (2010). *Science and Religion.* Malden: Wiley-Blackwell.
Stewart, W. Christopher. (2009). "On Giving Intelligent Design Theorists What They Say They Want" in *Divine Grace and Emerging Creation: Wesleyan Forays in Science and Theology of Creation*, edited by Thomas Oord. Eugene: Pickwick.
Stone, Bryan P. and Thomas Jay Oord, editors. (2001). *Thy Nature and Thy Name Is Love: Wesleyan and Process Theologies in Dialogue.* Nashville: Abingdon.
Stone, Ronald H. (2001). *John Wesley's Life and Ethics.* Nashville: Abingdon.
Sturm, Douglas. (1988). *Community and Alienation: Essays on Process Thought and Public Life.* Notre Dame: University of Notre Dame.
Sturm, Douglas. (1998). *Solidarity and Suffering: Toward a Politics of Relationality.* Albany: State University of New York Press.

Sullivan, Woodruff T., III, editor. (1984). *The Early Years of Radio Astronomy: Reflections Fifty Years after Jansky*. Cambridge: Cambridge University Press.

Sullivan, Woodruff T., III (1990). "The Entry of Radio Astronomy into Cosmology: Radio Stars and Martin Ryle's 2C Survey" in *Modern Cosmology in Retrospect*, edited by R. Bertotti and others. Cambridge: Cambridge University Press.

Sullivan, Woodruff T., III (1991). "Some Highlights of Interferometry in Early Radio Astronomy" in *Radio Interferometry: Theory, Techniques and Applications, Proceedings of the 131st IAU Colloquium, ASP Conference Series, San Francisco, Volume 19*, edited by T. J. Cornwell and R. A. Perley.

Susskind, Leonard. (2006). *The Cosmic Landscape: String Theory and the Illusion of Intelligent Design*. New York: Little, Brown.

Susskind, Leonard. (January 26, 2009). "Cosmology Lecture 3," online at www.youtube.com/Stanford.

Swimme, Brian. (1988). "The Cosmic Creation Story" in *The Reenchantment of Science: Postmodern Proposals*, edited by David Ray Griffin. Albany: State University of New York Press.

Swimme, Brian and Thomas Berry. (1992). *The Universe Story*. San Francisco: Harper.

Swimme, Brian Thomas and Mary Evelyn Tucker. (2011). *Journey of the Universe*. New Haven: Yale University Press.

Sylvest, Edwin S. (Summer 1992). "Cosmic Race and Cosmic Grace: New Possibilities for Humankind" in *Apuntes*, vol. 12, issue 2.

Teerikorpi, Pekka and others (2009). *The Evolving Universe and the Origin of Life: The Search for Our Cosmic Roots*. New York: Springer.

Temple, Robert. (2007). *The Genius of China: 3,000 Years of Science, Discovery, and Invention*, revised edition, foreword by Joseph Needham. Rochester: Inner Traditions [originally, New York: Simon and Schuster, 1986].

Templeton, John Marks. (1981). *The Humble Approach: Scientists Discover God*. New York: Seabury; revised, Continuum, 1995.

Templeton, John Marks. (1997). *How Large Is God?: Voices of Scientists and Theologians*. Philadelphia: Templeton.

Templeton, John Marks. (2000). *Possibilities for Over One Hundredfold More Spiritual Information: The Humble Approach in Theology and Science*. Philadelphia: Templeton.

Templeton, John Marks, editor. (1994). *Evidence of Purpose: Scientists Discover the Creator*. New York: Continuum; Templeton, 1997.

Templeton, John Marks and Robert L. Herrmann. (1994). *Is God the Only Reality?: Science Points to a Deeper Meaning of the Universe*. New York: Continuum.

Templeton, John Marks and Robert L. Herrmann. (1998). *The God Who Would Be Known: Revelations of the Divine in Contemporary Science*. Philadelphia: Templeton.

Tepfer, David and Sydney Leach. (December [June 22] 2006). "Plant Seeds as Model Vectors for the Transfer of Life through Space" in *Astrophysics and Space Science*, vol. 306, issue 1–2, 69–75 [doi:10.1007/s10509–006–9239–0].

Teplitz, H. I. with V. Desai and others. (2007). "Measuring PAH Emission in Ultradeep Spitzer IRS Spectroscopy of High-Red-Shift IR-Luminous Galaxies" in *The Astrophysical Journal*, vol. 659, issue 2, 941–49.

Thirring, Walter E. (c. 2007). *Cosmic Impressions: Traces of God in the Laws of Nature*. Philadelphia: Templeton.

Thomas, Hugh. (1997). *The Slave Trade: The Story of the Atlantic Slave Trade: 1440–1870*. New York: Simon and Schuster.

Thomas, Paul J. with Christopher F. Chyba and Christopher P. McKay, editors. (1997). *Comets and the Origin and Evolution of Life*. New York: Springer; Berlin: Springer-Verlag, 2006.

Tiberius, Valerie. (2015). *Moral Psychology: A Contemporary Introduction*. New York: Routledge.

Toulmin, Stephen and June Goodfield (1962). *The Architecture of Matter*. New York: Harper.

Toulmin, Stephen and June Goodfield. (1977). *The Discovery of Time*. Chicago: University of Chicago Press.

Towne, Edgar A. (May 2001). "New Physics and Hartshorne's Dipolar Theism" in *American Journal of Theology and Philosophy*, vol. 22, issue 2, 114–32.

Treschow, Michael with Willemien Otten and Walter Hannam, editors. (2007). *Divine Creation in Ancient, Medieval, and Early Modern Thought: Essays Presented to the Rev'd Dr. Robert D. Crouse*. Leiden: Boston.

Tu, Weiming. (1979). *Humanity and Self-Cultivation: Essays in Confucian Thought*. Berkeley: Asian Humanities.

Tu, Weiming. (1989 [1976]). *Centrality and Community: An Essay on Confucian Religiousness*, revised and enlarged. Albany: State University of New York Press.

Tunstall, Dwayne A. (2013). *Doing Philosophy Personally: Thinking about Metaphysics, Theism, and Antiblack Racism*. Bronx: Fordham University Press.

Turman, Eboni Marshall (2013). *Toward a Womanist Ethic of Incarnation: Black Bodies, the Black Church, and the Council of Chalcedon*. New York: Palgrave.

Turner, Michael S. (September 2009). "Origin of the Universe" in *Scientific American*, vol. 301, 36–43 [doi:10.1038/scientificamerican0909–36].

Tuszynski, Jack A. (March 2014). "Need for a Physical Basis of Cognitive Process: Comment on 'Consciousness in the Universe: A Review of the 'Orch OR' Theory' by Hameroff and Penrose" in *Physics of Life Reviews*, vol. 11, issue 1, 79–80.

Tymieniecka, Anna Teresa. (2009). *The Fullness of the Logos in the Key of Life: Book One, The Case of God in the New Enlightenment*. New York: Springer.

Tymieniecka, Anna Teresa. (2012). *Phenomenology and the Human Positioning in the Cosmos: The Life-World, Nature, Earth: Book One*. New York: Springer.

Tymieniecka, Anna Teresa. (2014). *Phenomenology of Space and Time: Book 2, The Forces of the Cosmos and the Ontopoietic Genesis of Life*. New York: Springer.

Tymieniecka, Anna-Teresa and Attila Grandpierre, editors. (2011). *Astronomy and Civilization in the New Enlightenment: Passions of the Skies [Analecta Husserliana: The Yearbook of Phenomenological Research, Volume CVII]*. Dordrecht: Springer.

Tyson, Neil de Grasse and Donald Goldsmith. (2004). *Origins: Fourteen Billion Years of Cosmic Evolution*. New York: Norton.

Unger, Roberto. (2014). *The Religion of the Future*. Cambridge: Cambridge University Press.

Unger, Roberto Mangabeira and Lee Smolin. (2015). *The Singular Universe and the Reality of Time: A Proposal in Natural Philosophy*. Cambridge: Cambridge University Press.

Urban, Wilbur Marshall. (1941). "Whitehead's Philosophy of Language and its Relation to His Metaphysics" in *The Philosophy of Alfred North Whitehead, The Library of Living Philosophers, Vol. 3*, edited by P. A. Schilpp. La Salle: Open Court.

Urey, Harold Clayton. (1952). *The Planets, Their Origin and Development*. New Haven: Yale University Press.

Urey, Harold C. (March 24, 1962). "Life-Forms in Meteorites: Origin of Life-like Forms in Carbonaceous Chondrites Introduction" in *Nature*, vol. 193, issue 4821, 1119–23 [doi:10.1038/1931119a0].

Urey, Harold Clayton. (1963). *Some Cosmochemical Problems*. University Park: Pennsylvania State University.

Vakoch, Douglas A., editor. (2014). *Extraterrestrial Altruism: Evolution and Ethics in the Cosmos*. Berlin: Springer.

Vân, Jean Trân Thanh with Yannick Mellier and Marc Moniez, editors. (c. 2001). *Cosmological Physics with Gravitational Lensing: Proceedings of the XXXVth Rencontres de Moriond*. Moriond Astrophysics Meetings, March 11–18, 2000, Les Arcs, Savoie, France. Les Ulis: EDP Sciences.

Van Huyssteen, J. Wentzel. (1998). *Duet or Duel?: Theology and Science in a Postmodern World*. Harrisburg: Trinity.

Van Huyssteen, J. Wentzel. (2006). *Alone in the World?: Human Uniqueness in Science and Theology*. Grand Rapids: Eerdmans.

Van Huyssteen, J. Wentzel and Khalil Chamcham, editors. (2012). *Templeton Science and Religion Reader*. West Conshohocken: Templeton.

Van Nuys, Kelvin. (1949). *Science and Cosmic Purpose*. New York: Harper.

Vanysek, V. and N. C. Wickramasinghe. (October 1999 [April 1975]). "Formaldehyde Polymers in Comets" in *Astrophysics and Space Science*, vol. 268, issue 1–3, 115–24 [originally, vol. 33, issue 2, L19–L28] [doi:10.1023/A:1002452904168].

Veneziano, Gabriele. (February 1, 2012). "The Myth of the Beginning of Time: Two Views of the Beginning" in *Scientific American*, vol. 21, 78–89 [doi:10.1038/scientificamericantime0212–78].

Venter, J. Craig with Mark D. Adams and others. (2001). "The Sequence of the Human Genome" in *Science*, vol. 291, issue 5507, 1304–51.

Vilenkin, Alexander. (November 4 [June 11], 1982). "Creation of Universes from Nothing" in *Physical Letters*, vol. 117B, issue 1, 2, 25–28.

Vilenkin, Alexander (1985). "Cosmic Strings and Domain Walls" in *Physics Reports*, vol. 121, issue 5, 263–315 [doi:10.1016/0370–1573(85)90033-X].

Vilenkin, Alexander. (2006). *Many Worlds in One: The Search for Other Universes*. New York: Hill and Wang.

Viney, Donald Wayne. (1985). *Charles Hartshorne and the Existence of God*. Albany: State University of New York Press.

Viney, Donald Wayne. (May 2001). "Is the Divine Shorn of Its Heart? Responding to Simoni-Wastila" in *American Journal of Theology and Philosophy*, vol. 22, issue 2, 155–72.

Viney, Donald Wayne. (Spring 2014). "Process Theism" in *Stanford Encyclopedia of Philosophy*, edited by E. N. Zalta, online at http://plato.standored.edu/archieves/apr2014/entries/process-theism/.

Vishio, Alexander. (2006). "Revisionary Neoclassical Theism." University of Chicago, unpublished.

Voskuil, Duane. (Fall–Winter 1999). "Hartshorne, God and Metaphysics: How the Cosmically Inclusive Personal Nexus and the World Interact" in *Process Studies*, vol. 28, issue 3/4, 212–28.

Vsekhsvyatskiy, Sergei Konstantinovich. (1970). *The Nature and Origin of Comets and Meteors*. Washington: NASA.

Wagoner, Robert V. with William A. Fowler and Fred Hoyle. (April 1967 [September 1, 1966]). "On the Synthesis of Elements at Very High Temperatures" in *The Astrophysical Journal*, vol. 148, issue 1, 3–49 [doi:10.1086/149126].

Wainwright, Milton. (December 17, 2014). "Astrobiology: Prescient Words on Comets and Life" in *Nature*, vol. 516, 329 [doi:10.1038/516329a].

Wainwright, Milton with N. C. Wickramasinghe, J. V. Narlikar, and P. Rajaratnam. (January 2003 [September 24, October 31, November 5, December 3, 2002]). "Microorganisms Cultured from Stratospheric Air Samples Obtained at 41km" in *Federation of European Microbiological Societies—Microbiology Letters*, vol. 218, issue 1, 161–65.

Wainwright, Milton with Fawaz Alshammari and Khalid Alabri. (May 2010). "Are Microbes Currently Arriving to Earth from Space?" in *Journal of Cosmology*, vol. 7, issue 3, 1692–702.

Wainwright, Milton and others. (2013). "Biological Entities Isolated from the Stratosphere (22–27km): Case for Their Space Origin" in *Instruments, Methods, and Missions for Astrobiology, XVI, Proceedings of SPIE*, vol. 8865.

Wainwright, Milton with Christopher E. Rose, Alexander J. Baker, and N. Chandra Wickramasinghe. (January 2014). "Impact Events on a Graphite Stub Provide Evidence that a Biological Entity Arrived to the Stratosphere from Space" in *Journal of Cosmology*, vol. 23, issue 7, 11131–35.

Waldrop, M. Mitchell. (1992). *Complexity: The Emerging Science at the Edge of Order and Chaos*. New York: Simon & Schuster.

Waldrop, M. Mitchell. (February 16, 2011). "Religion: Faith in Science" [concerning Templeton Foundation as "friend of science"] in *Nature*, vol. 470, 323–25 [doi:10.1038/470323a].

Walker, Christopher, editor. (1996). *Astronomy before the Telescope*, foreword by Patrick Moore. New York: St. Martin's.

Walker, Erica N. (2014). *Beyond Banneker: Black Mathematicians and the Paths to Excellence*. Albany: State University of New York Press.

Walker, Theodore Jr. (2004). *Mothership Connections: A Black Atlantic Synthesis of Neoclassical Metaphysics and Black Theology*. Albany: State University of New York Press.

Walker, Theodore Jr. (December 2006). "A Review of Stephen W. Hawking and Leonard Mlodinow's *A Briefer History of Time* (2005)" in *Journal of the American Academy of Religion*, vol. 74, issue 4, 1037–39.

Walker, Theodore Jr. (2008). "Neoclassical Cosmology and Matthew 22:36–40: Necessary Cosmological Distinctions Imply Theological Ethics" in *Whiteheadian Ethics: Abstracts and Papers from the Ethics Section of the Philosophy Group at the 6th International Whitehead Conference at the University of Salzburg, July 2006*, edited by Theodore Walker Jr. and Mihály Tóth. Newcastle: Cambridge Scholars.

Walker, Theodore Jr. (April 2010a). "Important Distinctions in Neoclassical Cosmology and Joseph's 'The Infinite Universe vs the Myth of the Big Bang'" in *Journal of Cosmology*, vol. 6, issue 13, 1616–20.

Walker, Theodore Jr. (2010b). "'Yes!' to Natural Theology and Justice: Stanley Hauerwas, Martin Luther King Jr., and Charles Hartshorne" in *Unsettling Arguments: A Festschrift on the Occasion of Stanley Hauerwas's 70th Birthday*, edited by Charles R. Pinches, Kelly S. Johnson, and Charles M. Collier. Eugene: Cascade.

Walker, Theodore Jr. (September–October 2011). "Anthropic Reasoning about Fine-Tuning, and Neoclassical Cosmology: Providence, Omnipresence, and Observation Selection Theory" in *Journal of Cosmology*, vol. 16, issue 26, 6776–98.

Walker, Theodore Jr. (June 2012a). "Astro-Theology in the *Journal of Cosmology*" in *Journal of Cosmology*, vol. 19, issue 2, 8552–55.

Walker, Theodore Jr. (June 2012b). "The Liberating Role of Astronomy in an Old Farmer's Almanac: David Rittenhouse's 'Useful Knowledge' and a Benjamin Banneker Almanac for 1792" in *Journal of Cosmology*, vol. 19, issue 3, 8556–82.

Walker, Theodore Jr. (September 2012). "A Review of Carl Sagan's *The Varieties of Scientific Experience*, edited by Ann Druyan (2006 [originally "Search for Who We Are," unpublished 1985])" in *Journal of Cosmology*, vol. 20, 8730–32.

Walker, Theodore Jr. (March 2013). "Postmodern Astro-Theology, Cometary Panspermia, and the Polonnaruwa Meteorite: Derham, Wesley, Whitehead, Griffin and Cobb" in *Journal of Cosmology*, vol. 22, issue 21, 10141–51.

Wallis, Jamie with Nori Miyake, Richard B. Hoover, Andrew Oldroyd, Daryl H. Wallis, Anil Samaranayake, K. Wickramarathne, M. K. Wallis, Carl H. Gibson, and Nalin Chandra Wickramasinghe. (March 5, 2013). "Polonnaruwa Meteorite: Oxygen Isotope, Crystalline and Biological Composition" in *Journal of Cosmology*, vol. 22, issue 2, 10004–11.

Wallis, Max K. (April 3, 1980a). "Radiogenic Melting of Primordial Comet Interiors" in *Nature*, vol. 284, issue 5755, 431–33 [doi:10.1038/284431a0].

Wallis, Max K. (July 17, 1980b). "Cometary Science" in *Nature*, vol. 286, issue 5770, 207–8 [doi:10.1038/286207a0].

Wallis, Max K. and N. C. Wickramasinghe, F. Hoyle, and R. Rabilizirov. (January 1989 [December 29, 1988]). "Biologic versus Abiotic Models of Cometary Grains" in *Monthly Notices of the Royal Astronomical Society*, vol. 238, 1165–70.

Wallis, Max K. and N. C. Wickramasinghe. (April 1991 [July 17, 1990]). "Structural Evolution of Cometary Surfaces" in *Space Science Reviews*, vol. 56, 93–97 [doi:10.1007/BF00178395].

Wallis, Max K. and N. C. Wickramasinghe. (1992). "Cometary Habitats for Primitive Life" in *Advances in Space Research*, vol. 12, issue 4, 281–85.

Wallis, Max K. and N. C. Wickramasinghe. (1995). "Role of Major Terrestrial Cratering Events in Dispersing Life in the Solar System" in *Earth and Planetary Science Letters*, vol. 130, 69–73.

Wallis, Max K. and Sirwan Al-Mufti. (February 1996). "Processing of Cometary Grains at the Nucleus Surface" in *Earth, Moon, and Planets*, vol. 72, issue 1–3, 91–97 [doi:10.1007/BF00117507].

Wallis, Max K. and Nalin Chandra Wickramasinghe. (2004). "Interstellar Transfer of Planetary Microbiota" in *Monthly Notices of the Royal Astronomical Society*, vol. 348, issue 1, 52–57, 61.

Wan, Sze-kar. (December [October 15] 2008). "The Viability of Confucian Transcendence: Grappling with Tu Weiming's Interpretation of the *Zhongyong*" in *Dao*, vol. 7, issue 4, 407–21 [doi:10.1007/s11712-008-9089-2].

Wang, Quian and Chris Tourney. (October 24, 2011). "Long History versus Good Science" in *Nature Chemistry*, vol. 3, issue 11, 832–33 [doi:10.1038/nchem.1179].

Wang, Xiaoxia with Nivedita Mitra, Ismael Secundino, and others. (June 19, 2012). "Specific Inactivation of Two Immunomodulatory SIGLEC Genes during Human Evolution" in *Proceedings of the National Academy of Science*, vol. 109, issue 25, 9935–40 [doi:10.1073/pnas.1119459109].

Ward, Keith. (2004). "World as the Body of God: A Panentheistic Metaphor" in *In Whom We Live and Move and Have Our Being: Panentheistic Reflections on God's Presence in a Scientific World*, edited by P. Clayton and A. Peacocke. Grand Rapids: Eerdmans.

Ward, Keith. (2007). *Divine Action: Examining God's Role in an Open and Emergent Universe*. Philadelphia: Templeton.

Ward, Keith. (2008a). *The Big Questions in Science and Religion*. West Conshohocken: Templeton.

Ward, Keith. (2008b). *Why There Almost Certainly Is a God: Doubting Dawkins.* Oxford: Lion Hudson.

Warren, Richard. (2002). *The Purpose Driven Life.* Grand Rapids: Zondervan/HarperCollins.

Wassserburg, G. J. with William A. Fowler and Fred Hoyle. (February [January 31] 1960). "Duration of Nucleosynthesis" in *Physical Review Letters*, vol. 4, issue 3, 112–14 [doi:10.1103/PhysRevLett.4.112].

Watts, Fraser. (2013). "Natural Theology and the Mind Sciences" in *The Oxford Handbook of Natural Theology*, edited by Russell Re Manning. Oxford: Oxford University Press.

Watts, Fraser and Kevin Dutton, editors. (2006). *Why the Science and Religion Dialogue Matters: Voices from the International Society for Science and Religion.* Philadelphia: Templeton.

Weinberg, Steven. (1972). *Gravitation and Cosmology: Principles and Applications of the General Theory Relativity.* New York: Wiley.

Weinberg, Steven. (January 1989). "The Cosmological Constant Problem" in *Reviews of Modern Physics*, vol. 61, issue 1, 1–23 [doi:10.1103/RevModPhys.61.1].

Weinberg, Steven. (1993 [1977]). *The First Three Minutes: A Modern View of the Origin of the Universe*, updated. New York: Basic.

Weinberg, Steven. (October 1994). "Life in the Universe" in *Scientific American*, vol. 271, issue 4, 44–49.

Weinberg, Steven. (2008). *Cosmology.* Oxford: Oxford University Press.

Weinberg, Steven. (2015). *To Explain the World: The Discovery of Modern Science.* New York: Harper.

Wesley, John. (1784). *A Survey of the Wisdom of God in Creation: A Compendium of Natural Philosophy*, fourth edition. London: Printed for J. Paramore Upper-Moorfields.

Westaway, Frederic William (1920). *Science and Theology: Their Common Aims and Methods.* London: Blackie Publishing.

Westerhoff, Jan. (2011). *Reality: A Very Short Introduction.* Oxford: Oxford University Press.

White, Morton, editor. (1955). *The Age of Analysis: 20th Century Philosophers.* New York: Mentor.

White, S. D. M. and M. J. Rees. (May 1978 [September 26, 1977]). "Core Condensation in Heavy Halos: A Two-Stage Theory for Galaxy Formation and Clustering" in *Monthly Notices of the Royal Astronomical Society*, vol. 183, 341–58.

Whitehead, Alfred North. (1919). *An Enquiry Concerning the Principles of Natural Knowledge.* Cambridge: Cambridge University Press.

Whitehead, Alfred North. (November 15, 1919). "A Revolution in Science" in *Nation*, vol. 26, 232–33 [according to Whitehead bibliography in *The Philosophy of Alfred North Whitehead* (1971) edited by P. A. Schilpp].

Whitehead, Alfred North. (1920). *The Concept of Nature* (Tarner Lectures, 1919). Cambridge: Cambridge University Press.

Whitehead, Alfred North. (1922). *The Principle of Relativity with Applications to Physical Science.* Cambridge: Cambridge University Press; Mineola: Dover, 2004.

Whitehead, Alfred North. (1925). *Science and the Modern World* (Lowell Lectures, 1925). New York: Macmillan; Free Press, 1967.

Whitehead, Alfred North. (1926). *Religion in the Making* (Lowell Lectures, 1926). New York: Macmillan.

Whitehead, Alfred North. (1926). "Time" in *Proceedings of the Sixth International Congress of Philosophy,* September 13–17, Harvard University, edited by E. S. Brightman. New York: Longmans, Green.

Whitehead, Alfred North. (1927). *Symbolism, Its Meaning and Effect.* New York: Macmillan.

Whitehead, Alfred North. (1927–28). *Process and Reality: An Essay in Cosmology* (Gifford Lectures), *1978 Corrected Edition,* edited by David Ray Griffin and Donald W. Sherburne. New York: Free Press.

Whitehead, Alfred North. (1929). "The Function of Reason." Louis Clark Vanuxem Foundation Lectures delivered at Princeton University. Princeton: Princeton University Press; Boston: Beacon, 1958.

Whitehead, Alfred North. (1933). *Adventures of Ideas.* New York: Macmillan.

Whitehead, Alfred North. (1934). *Nature and Life.* Chicago: University of Chicago Press.

Whitehead, Alfred North. (1938). *Modes of Thought.* Six lectures delivered at Wellesley College, Massachusetts [1937–38] and two lectures at the University of Chicago [1933]. New York: Macmillan.

Whitehead, Alfred North. (1941). "Autobiographical Notes" [1–14], "Mathematics and the Good" [666–81], and "Immortality" [682–700] in *The Philosophy of Alfred North Whitehead, The Library of Living Philosophers, Vol. 3,* edited by Paul Arthur Schilpp. La Salle: Open Court.

Whitehead, Alfred North. (1947). *Essays in Science and Philosophy.* New York: Philosophical Library.

Whitehead, Alfred North and Bertrand Russell. (1910, 1912, 1913). *Principia Mathematica, Vols. 1, 2, and 3.* Cambridge: Cambridge University Press.

Whittet, D. C. B. (October 25, 1979). "Interstellar Grains: Organic or Refractory?" in *Nature,* vol. 281, issue 5733, 708 [doi:10.1038/281708a0].

Wickramasinghe, Dayal Tissa and David A. Allen. (October 9, 1980). "The 3.4-μm Interstellar Absorption Feature" in *Nature,* vol. 287, issue 5782, 518–19 [doi:10.1038/287518a0].

Wickramasinghe, D. T. and D. A. Allen. (September 4, 1986). "Discovery of Organic Grains in Comet Halley" in *Nature,* vol. 323, issue 6083, 44–46 [doi:10.1038/323044a0].

Wickramasinghe, D. T. and D. A. Allen. (October 21, 1987). "Discovery of Organic Grains in Comet Wilson" in *Nature,* vol. 329, issue 6140, 615–16 [doi:10.1038/329615a0].

Wickramasinghe, D. T. with F. Hoyle, N. C. Wickramasinghe, and S. Al-Mufti. (June 24, 1986). "A Model of the 2–4 Micron Spectrum of Comet Halley" in *Earth, Moon, and Planets,* vol. 36, 295–99.

Wickramasinghe, Janaki Tara. (May 25, 2007). *The Role of Comets in Panspermia.* Cardiff University PhD Thesis. Ann Arbor: ProQuest LLC Publishing, 2013 [MUI Dissertation Publishing, UMI Number: U584951].

Wickramasinghe, Janaki Tara and N. C. Wickramasinghe. (December 2006). "A Cosmic Prevalence of Nanobacteria?" in *Astrophysics and Space Science*, vol. 305, issue 4, 411–13 [doi:10.1007/s10509–006–9181–1].

Wickramasinghe, Janaki Tara with Chandra Wickramasinghe and William Napier. (2010). *Comets and the Origin of Life.* Hackensack: World Scientific.

Wickramasinghe, Nalin Chandra. (June 13, 1962). "On Graphite Particles as Interstellar Grains" in *Monthly Notices of the Royal Astronomical Society*, vol. 124, issue 5, 417–33.

Wickramasinghe, N. C. (February 8, 1963 [November 8, 1962]). "On Graphite Particles as Interstellar Grains, II" in *Monthly Notices of the Royal Astronomical Society*, vol. 126, issue 1, 7, 99–114.

Wickramasinghe, N. C. (January 4, 1965). "On the Growth and Destruction of Ice Mantles on Interstellar Graphite Grains" in *Monthly Notices of the Royal Astronomical Society*, vol. 131, issue 2, 177–90.

Wickramasinghe, N. C. (1966 [January 7, 1965]). "On the Optics of Small Graphite Spheres, I" in *Monthly Notices of the Royal Astronomical Society*, vol. 131, issue 3, 18, 263–69.

Wickramasinghe, N. Chandra. (1967). *Interstellar Grains.* London: Chapman & Hall.

Wickramasinghe, N. C. (November 15, 1969). "Interstellar Polarization by Graphite-Silicate Grain Mixtures" in *Nature*, vol. 224, issue 5220, 656–58 [doi:10.1038/224656a0].

Wickramasinghe, N. C. (October 31, 1970). "Between the Stars" in *Nature*, vol. 228, issue 5270, 483–84 [doi:10.1038/228483c0].

Wickramasinghe, N. C. (December 6, 1974). "Formaldehyde Polymers in Inter stellar Space" in *Nature*, vol. 252, issue 5483, 462–63 [doi:10.1038 /252462a0].

Wickramasinghe, Chandra. (1980). "The Origin of Life on Earth," audio lecture, online at www.buckingham.ac.uk/research/bcab/hrwarchieve.

Wickramasinghe, Chandra. (2001). *Cosmic Dragons: Life and Death on Our Planet.* London: Souvenir Press.

Wickramasinghe, Chandra. (2004 [September 29, 2003]). "The Universe: A Cryogenic Habitat for Microbial Life" in *Cryobiology*, vol. 48, 113–25.

Wickramasinghe, Chandra. (2005a). "Alternative Cosmologies" in *A Journey with Fred Hoyle: The Search for Cosmic Life*, edited by Kamala Wickramasinghe. London: World Scientific.

Wickramasinghe, Chandra. (2005b). "From Dust to Life" in *The Scientific Legacy of Fred Hoyle*, edited by D. Gough, foreword by M. Rees. Cambridge: Cambridge University Press, 2011 paperback.

Wickramasinghe, Chandra. (August 15, 2007). "Rosetta Mission to a Comet: Our Involvement and Interest in Rosetta," online at www.cardiffcentreforastrogiol ogy.com/Rosetta.html [accessed April 14, 2014].

Wickramasinghe, Chandra. (April [January 29] 2010). "Astrobiological Case for Our Cosmic Ancestry" in *International Journal of Astrobiology*, vol. 9, issue 2, 119–29 [doi:10.1017/s1473550409990413].

Wickramasinghe, Chandra. (November 4, 2010). "Microfossils in Comet Dust and Meteorites Support Panspermia" in *Proceedings of SPIE*[doi:10.1117/2.1201010.003239].

Wickramasinghe, Chandra. (2011). "The Compelling Case for Panspermia" in *Astronomy and Civilization in the New Enlightenment: Passions of the Skies* [*Analecta Husserliana: The Yearbook of Phenomenological Research, Volume CVII*], edited by Anna-Teresa Tymieniecka and Attila Grandpierre. Dordrecht: Springer.

Wickramasinghe, Chandra. (September–October 2011). "From Astrochemistry to Astrobiology" in *Journal of Cosmology*, vol. 16, issue 1, 6519–27.

Wickramasinghe, Chandra. (December 2011). "A Galaxy Strewn with Microorganisms" in *A Destiny of Cosmic Life: Chapters in the Life of an Astrobiologist*. Kindle/Amazon Books, released January 2, 2014.

Wickramasinghe, N. Chandra. (January 2013). "DNA Sequencing and Predictions of the Cosmic Theory of Life" in *Astrophysics and Space Science*, vol. 343, issue 1, 1–5 [doi:10.1007/s10509–012–1227-y].

Wickramasinghe, N. C. (February 2013). "Simulation of Earth-based Theory with Lifeless Results" [review of *First Life: Discovering the Connections between Stars, Cells, and How Life Began* (2012) by David Deamer] in *Bio Science*, vol. 63, issue 2, 141–43.

Wickramasinghe, Chandra. (2014). *The Search for Our Cosmic Ancestry*. Hackensack: World Scientific.

Wickramasinghe, N. C. (July 24 [July 16], 2014). "ROSETTA's Prospects for Discovering Evidence of Cometary Biology" in *Journal of Astrobiology and Outreach*, vol. 2, issue 3, 1–3/118 [doi:10.4172/2332–2519.1000118].

Wickramasinghe, N. Chandra. (September 2014). "Comet 67P/Churyumov-Gerasimenko and Cometary Biology" in *Journal of Cosmology*, vol. 24, issue 3, 12032–36.

Wickramasinghe, Chandra. (2015). *Where Did We Come From?: Life of an Astrobiologist*, edited by Kamala Wickramasinghe. Hackensack: World Scientific.

Wickramasinghe, N. C. and C. Guillaume (July 24, 1965). "Interstellar Extinction by Graphite Grains" in *Nature*, vol. 207, issue 4995, 366–68 [doi:10.1038/207366a0].

Wickramasinghe, N. C. and J. V. Narlikar. (October 7, 1967). "Microwave Background in a Steady-State Universe" in *Nature*, vol. 216, issue 5110, 43–44 [doi:10.1038/216043a0].

Wickramasinghe, N. C. and K. S. Krishna Swamy. (1968 [November 13, 1967]). "On the Temperature of Interstellar Grains" in *Monthly Notices of the Royal Astronomical Society*, vol. 139, issue 3, 283–87.

Wickramasinghe, N. C. with J. G. Ireland, K. Nandy, H. Seddon, and R. D. Wolstencroft. (February 3, 1968). "Origin of the Diffuse Interstellar Bands" in *Nature*, vol. 217, issue 5127, 412–15 [doi:10.1038/217412b0].

Wickramasinghe, N. C. with F. Hoyle and K. Nandy. (1977). "Organic Molecules in Interstellar Dust: A Possible Spectral Signature at λ2200 Å?" in *Astrophysics and Space Science* vol. 268, issue 1–3, 295–99 [vol. 47, L9–L11].

Wickramasinghe, N. C. with F. Hoyle, J. Brooks, and G. Shaw. (October 20, 1977). "Prebiotic Polymers and Infrared Spectra of Galactic Sources" in *Nature*, vol. 269, issue 5630, 674–76 [doi:10.1038/269674a0].

Wickramasinghe, N. C. and A. H. Olavesen. (October 26, 1978). "Cosmochemistry and Evolution" in *Nature*, vol. 275, issue 5682, 694 [doi:10.1038/275694a0].

Wickramasinghe, N. C. with A. N. Wickramasinghe and Fred Hoyle. (October 1992). "The Case against Graphite Particles in Interstellar Space" in *Astrophysics and Space Science*, vol. 196, issue 1, 167–69 [doi:10.1007/BF00645249].

Wickramasinghe, Chandra with F. Hoyle and D. Lloyd. (1996). "Eruptions of Comet Hale-Bopp at 6.5 AU" in *Astrophysics and Space Science*, vol. 240, 161–65.

Wickramasinghe, Nalin C. with Fred Hoyle, Shirwan Mufti, and Daryl H. Wallis. (January 1997 [July 1–5, 1996]). "Infrared Signatures of Prebiology or Biology?" in *Astronomical and Biochemical Origins and the Search for Life in the Universe; IAU Colloquium*, edited by Cristiano Batalli Cosmovici, Stuart Bowyer, and Dan Werthimer. Bologna: Editrice Compositori.

Wickramasinghe, N. C. and Daisaku Ikeda. (1998). *Space and Eternal Life: A Dialogue between Chandra Wickramasinghe and Daisaku Ikeda*, foreword by Fred Hoyle. Chicago: Journeyman.

Wickramasinghe, N. C. with J. V. Narlikar, J. T. Wickramasinghe, and M. Wainwright. (2003). "Expanding Horizons of Cosmic Life" in *Proceedings of SPIE*, vol. 4859, 154–63.

Wickramasinghe, Chandra with Jayant Narlikar and Geoffrey Burbidge, editors. (2003). *Fred Hoyle's Universe*. Dordrecht: Kluwer [from *Proceedings of a Conference Celebrating Fred Hoyle's Extraordinary Contributions to Science*, June 25–26, 2002, Cardiff University].

Wickramasinghe, N. Chandra and Carl H. Gibson. (March 2 [February 27], 2010). "Imperatives of Cosmic Biology" in *Inspire: High Energy Physics Information System* [arXiv:1003.0091].

Wickramasinghe, N. Chandra and Janaki Tara Wickramasinghe. (Spring 2012 [June 5, 2012]). "A Note on Venus Transit and Microbial Injection to Earth" in *Journal of Cosmology*, vol. 18, issue 20, 8506–10.

Wickramasinghe, N. C. with J. Wallis, D. H. Wallis, and A. Samaranayake. (January 2013). "Fossil Diatoms in a New Carbonaceous Meteorite" in *Journal of Cosmology*, vol. 21, issue 37, 9560–71.

Wickramasinghe, N. C. with J. Wallis, D. H. Wallis, M. K. Wallis, S. Al-Mufti, J. T. Wickramasinghe, A. Samaranayake, and K. Wickramarathne. (January 2013). "On the Cometary Origin of the Polonnaruwa Meteorite" in *Journal of Cosmology*, vol. 21, issue 38, 9572–78.

Wickramasinghe, N. C. with J. Wallis, D. H. Wallis, M. K. Wallis, N. Miyake, S. G. Coulson, Carol H. Gibson, J. T. Wickramasinghe, A. Samaranayake, K. Wickramarathne, and Richard B. Hoover. (March 4, 2013). "Incidence of Low Density Meteoroids of the Polonnaruwa-Type" in *Journal of Cosmology*, vol. 22, issue 1, 1–8.

Wickramasinghe, N. Chandra and Gensuke Tokoro. (January 2014). "Life as a Cosmic Phenomenon: 2. The Panspermia Trajectory of *Homo sapiens*" in *Astrobiology and Outreach*, vol. 2, issue 2, 115 [doi:10.4172/2332–2519 .1000115].

Williams, Thomas, translator. (1995 [1977]). "Proslogion" (originally 1077–78) by Saint Anselm of Canterbury in *Classics of Western Philosophy*, fourth edition, edited by Steven M. Cahn. Indianapolis: Hackett.

Wilson, Edward O. (1984). *Biophilia: The Human Bond with Other Species*. Cambridge: Harvard University Press.

Wilson, Patrick A. (1993). "The Anthropic Principle" in *Cosmology: Historical, Literary, Philosophical, Religious, and Scientific Perspectives*, edited by N. S. Hetherington. New York: Garland; Routledge, 2008.

Witham, Larry. (2005). *The Measure of God: Our Century-Long Struggle to Reconcile Science and Religion*. New York: HarperCollins.

Witham, Larry. (2008). *The Proof of God: The Debate That Shaped Modern Belief*. New York: Atlas and Co.

Witze, Alexandra. (December 16, 2014). "Curiosity Rover Sniffs Martian Methane: Background Gas Levels Are Low—But They Spike Periodically" in *Nature News* [doi:10.1038/nature.2014.16578].

Wolfram, Stephen. (2002). *A New Kind of Science*. Champaign: Wolfram Media.

Wood, Charles M. (July 1999). "How Does God Act" in *International Journal of Systematic Theology*, vol. 1, issue 2, 138–52.

Wood, Charles M. (2008). "The Scope of Providence" in *The Question of Providence*. Louisville: Westminster John Knox.

Woosley, S. E. (1999). "Hoyle and Fowler's Nucleosynthesis in Supernovae" in *The Astrophysical Journal*, vol. 525, 924–25.

Woosley, S. E. (2007). "Nuclear Astrophysics: The First 50 Years" in *Nature Physics*, vol. 3, issue 12, 832–33 [doi:10.1038/nphys804].

Wouk, Herman. (2010). *The Language God Talks: On Science and Religion*. New York: Little Brown.

Wright, Thomas. (1971 [1750]). *An Original Theory or New Hypothesis of the Universe*, introduction and transcription by Michael A. Hoskin. New York: American Elsevier; London: Macdonald and Company [in facsimile reprint together with Thomas Wright's *A Theory of the Universe* (1734) via Library of Congress].

Young, Henry James. (1990). *Process: A Theology of Social Pluralism*. Minneapolis: Fortress.

Yourgrau, Wolfgang and Allen D. Breck, editors. (1970). *Physics, Logic, and History* [based on the First International Colloquium held at the University of Denver, Spring 1966]. New York: Plenum.

Yourgrau, Wolfgang and Allen D. Breck, editors. (1972). *Biology, History, and Natural Philosophy* [based on the Second International Colloquium held at the University of Denver, Fall 1967]. New York: Plenum.

Yourgrau, Wolfgang and Allen D. Breck, editors. (1977). *Cosmology, History, and Theology* [based on Third International Colloquium held at the University of Denver, Fall 1974]. New York: Plenum.

Zenil, Hector, editor. (2012 [Alan Turing Year]). *A Computable Universe: Understanding and Exploring Nature as Computation*, foreword by Roger Penrose. Hackensack: World Scientific.

About the Authors

Chandra Wickramasinghe (Buckingham Centre for Astrobiology, University of Buckingham, UK; Centre for Astrobiology, University of Peradeniya, Sri Lanka; editor in chief of the Journal of Astrobiology and Outreach) is an astronomer-astrobiologist and longtime research collaborator with Sir Fred Hoyle. Among other contributions to astronomy, astrobiology, and cosmology, they discovered organic molecules in interstellar space and advanced "cometary panspermia," the idea that water and organic matter are circulated by comets (a theory consistent with November 2014 data from the Rosetta space craft mission to Comet 67P/Churyumov-Gerasimenko).

Theodore Walker Jr. (Perkins School of Theology at Southern Methodist University in Dallas, Texas) is a liberty-oriented process theologian and theological ethicist. He serves on the International Advisory Board for the journal Process Studies, and he served as a co-editor with Joseph Bracken for the Astro-Theology volumes in the Journal of Cosmology.

Editing by Alexander F. Vishio.

Index

www.ingramcontent.com/pod-product-compliance
Lightning Source LLC
Chambersburg PA
CBHW061110140125
PP16098900001B/1
9781137552426